AF600285

WATER POLICY REFORM IN SOUTHERN ALBERTA

An Advocacy Coalition Approach

Water Policy Reform in Southern Alberta

An Advocacy Coalition Approach

B. TIMOTHY HEINMILLER

UNIVERSITY OF TORONTO PRESS
Toronto Buffalo London

Toronto Buffalo London
www.utppublishing.com

ISBN 978-1-4875-0053-5

Library and Archives Canada Cataloguing in Publication

Heinmiller, B. Timothy, 1975–, author
Water policy reform in southern Alberta : an advocacy coalition approach / B. Timothy Heinmiller.

Includes bibliographical references and index.
ISBN 978-1-4875-0053-5 (cloth)

1. Water-supply – Management – Government policy – Alberta.
2. Water conservation – Government policy – Alberta. I. Title.

TD227.A4H44 2016 333.910097123'4 C2016-903246-9

This book has been published with the help of a grant from the Federation for the Humanities and Social Sciences, through the Awards to Scholarly Publications Program, using funds provided by the Social Sciences and Humanities Research Council of Canada.

University of Toronto Press acknowledges the financial assistance to its publishing program of the Canada Council for the Arts and the Ontario Arts Council, an agency of the Government of Ontario.

Canada Council for the Arts Conseil des Arts du Canada

Funded by the Government of Canada Financé par le gouvernement du Canada

Contents

Acknowledgments

This project began its life during the year I spent as a postdoctoral fellow at the Economic Policy Research Institute at the University of Western Ontario in 2004–5. During this time, I benefited from the guidance of Drs Jim Davies, Andrew Sancton, and Robert Young, as well as the many faculty and students who provided feedback on my research presentations, the ideas for which eventually fed into this book. A special thanks also goes to Dr Robert Young and his Canada Research Chair for helping to fund some of my early research travel.

The bulk of the work for this book was undertaken during my time in the Political Science Department at Brock University, as an assistant/associate professor. Brock's Office of Research Services played an instrumental role in helping me to craft the application that secured the Social Sciences and Humanities Research Council (SSHRC) standard research grant that supported this research. Brock's research librarians, especially Heather Whipple and Colleen Beard, also provided invaluable and much-appreciated assistance in helping me track down obscure, out-of-province information and data sources. I also need to acknowledge the assistance of my departmental colleagues and my colleagues in the Environmental Sustainability Research Centre, with whom I explored ideas and discussed the finer points of my research methodology. Drs Matt Hennigar and Ryan Plummer deserve specific mention and thanks, in this regard. I also need to thank the plethora of Brock students, both graduate and undergraduate, who patiently listened to my self-indulgent research stories and were ever-ready with constructive, critical feedback on my work.

Key contributors to the development of this book, some of whom I have never met, were the librarians throughout Alberta who granted me access to their collections and/or arranged for inter-library loans. In particular, I owe a hearty thanks to the unknown librarian at the University of Alberta who consented to my inter-library loan request for over thirty volumes of transcripts from water hearings in the 1970s and 1980s. This single act of kindness made the entire research project possible. I'd also like to thank Doug Cass at the Glenbow Museum Library in Calgary for granting me access to their collection and for serving as a hospitable host during the week I spent there.

All research endeavours ultimately run on money, and I owe a great deal of thanks to those who funded this project. The largest funder was the SSHRC, whose funding afforded me the travel and student assistance that were essential to this project. Another key funder was the Environmental Sustainability Research Centre at Brock University, who provided funding to hire a second coder for the content-analysis portion of the research. This funding came at a critical moment in the project and is greatly appreciated.

I was fortunate to have excellent student research assistants on this project, all of them drawn from Brock's Master's Program in Political Science. Kristian Laughlin and Alexandra Swigger assisted on some of the early research, helping me to put Alberta's water policy reforms into context with what was happening in similar jurisdictions. Sydney Dick served as the intrepid second coder for the content analysis and did a fantastic job familiarizing herself with Alberta water issues and mastering the coding frame.

An extremely important group in my research were my interviewees. I am constantly amazed at the willingness of such busy and important individuals to make time for me and to share their policy development experiences. Without these many kindnesses, my research would have suffered greatly, and I would like to express my appreciation to each of you, even though I cannot name you individually.

As this book started to take shape, it was influenced by the feedback I received from papers presented at research conferences, including the annual general meetings of the Canadian Political Science Association and the Canadian Association of Programs in Public Administration. I also received very useful feedback on the research in chapter 6 from anonymous reviewers at the *Canadian Journal of Political Science*, and I thank them for that. Daniel Quinlan also provided useful and encouraging feedback on each chapter as they were being written, for which I must thank him.

Finally, I need to acknowledge the silent partner in this endeavour, my wife Ann. This book was a personal and professional odyssey for me and only she knows what went into its writing. I could not have completed it without her constant and unwavering support.

Oh, and I cannot forget Belle, the ninety-pound Bernese Mountain Dog who snored at my feet during the writing of the entire manuscript. I do not think I will ever read this book without associating it with her.

WATER POLICY REFORM IN SOUTHERN ALBERTA

An Advocacy Coalition Approach

Chapter One

Water Scarcity and Water Governance in Southern Alberta

Blessed with an abundance of riches in land, gas and oil, southern Alberta has encountered its first true geological limit: there simply isn't enough water to go around.

Renata D'Aliesio, 2007

Even in a province with a reputation for brashness, the project was unprecedented in size and ambition. It started with an announcement in October 2004. The United Horsemen of Alberta, having outgrown the racing facilities at the Calgary Stampede grounds, released plans for an $80-million racetrack, five-star hotel, and casino development in the small community of Balzac on the northern outskirts of Calgary. A year later, the United Horsemen were joined by the retail development firm Ivanhoe Cambridge, who purchased forty-nine hectares of land in the area and announced plans to construct a 1.4 million square foot shopping centre with around two hundred stores, including seventeen "big box" outlets. The mall itself would rival the well-known West Edmonton Mall in size and attraction, and the construction project would be the largest in Alberta outside of the northern oil sands. Altogether, the CrossIron project, as it became known, was a $1.5 billion investment into Alberta's already booming economy (D'Aliesio, "Part One" 2007).

The Municipal District of Rocky View, where CrossIron was situated, warmly welcomed the project, viewing it as a way to stimulate the local economy and increase the district's tax base. In relatively short order, Rocky View provided all of the necessary zoning and building approvals, and the project, scheduled for completion in 2007, was well underway. By late 2005, however, one hitch in the project started to emerge: water (D'Aliesio, "Part One" 2007).

Like most water users in southern Alberta, Rocky View had a water licence granted by the Alberta government. Since 1894, ownership of all water rights in the province had been vested in the Crown, and licences to use water had been controlled and distributed by government (Percy 1996–7). Most of Rocky View's water came from the nearby Bow River, but the district had already allocated just about all of its licensed water, leaving none for the new CrossIron development. The developers estimated that the new shopping centre needed about five million litres of water per day, a minute amount compared to the massive volumes used for irrigated agriculture in the region, but much more than Rocky View had available (D'Aliesio, "Part One" 2007).

Almost reflexively, the district applied to Alberta Environment for a new licence to take water from the Bow to support the CrossIron development. This followed a decades-long pattern in which water users looked to the Alberta government to supply water and water licences in support of new economic development, and the government, with few exceptions, duly complied (Glenn 1999). However, unbeknownst to Rocky View and the CrossIron developers, the window for obtaining new water licences in much of southern Alberta was quickly closing and they were about to be shut out. A new Water Act in 1999 had significantly reformed water governance in Alberta and, through the South Saskatchewan River Basin (SSRB) Water Management Plan completed in 2005, a decision had been made to impose moratoria on new water licences in three of southern Alberta's four sub-basins. The Oldman, South Saskatchewan, and Bow sub-basins were now closed to new licences, and all pending applicants, including Rocky View, would have to look elsewhere for water.

The only sub-basin not closed to new licences in the SSRB was the Red Deer sub-basin to the north of Rocky View, so it was there that they turned next. Rocky View made a second water licence application in June 2006, this time requesting water from the Red Deer, almost two hundred kilometres away. The application, which, under the terms of the Water Act, was posted for public comment, attracted a wave of protest from environmentalists and Red Deer communities concerned with protecting the source watershed. It also ignited controversy in the Alberta legislature as the opposition Liberals accused the Progressive Conservative (PC) government of secretly promising water for the CrossIron development – despite Rocky View's still pending licence application – after it was revealed that the developers had already secured an $8.3 million provincial grant for the project's water infrastructure. The public outcry and poor political optics combined to stall Rocky View's Red Deer licence application until the final nail in the Red Deer option was driven by the Town of Drumheller in May 2007. Drumheller, on the banks of the Red Deer River, was key to securing Red Deer water for Rocky View, because its water utilities were needed to treat and transport the water from the source. Responding to local concerns, the Drumheller Town Council twice voted against sending water to Rocky View, the second time in spite of a promise of $15 million from Rocky View for utility upgrades (D'Aliesio, "Part One" 2007).

As the Red Deer option was closing, Rocky View turned to its near neighbour, the City of Calgary. Unlike most communities in southern Alberta, Calgary had water to spare. With considerable forethought, the city had acquired water licences for enough water to support about 2.5 times its current population. Calgary was also practically on the doorstep of the CrossIron project, so it seemed to make sense for Rocky View to tap into Calgary's water system and take advantage of some of Calgary's surplus water. However, Calgary and Rocky View were also development rivals with a recent history of animosity, and it took the personal efforts of Municipal Affairs Minister Ray Danyluk to get the two parties to meet, on neutral ground, in Edmonton during March 2007. The City of Calgary, protecting its water to support future growth and wanting to discourage sprawling development on its fringes, offered to save the CrossIron project by annexing and servicing the Balzac area (and capturing its tax revenue), but refused to provide Rocky View with water. For Rocky View, annexation was a non-starter, and the talks soon reached an impasse. Rocky View might still find water, but it was not going to come from the City of Calgary (D'Aliesio, "Part One" 2007).

In addition to closing the Oldman, Bow, and South Saskatchewan sub-basins, the Water Act and SSRB Water Management Plan had, for the first time, introduced water licence trading. Water users in need of water, such as Rocky View, were now able to buy water licences from existing licence holders, if a willing seller could be found. Rocky View found such a seller in the Western Irrigation District (WID), located just east of Calgary. The WID is one of thirteen irrigation districts in southern Alberta, which hold most of the large water licences in the region and, moreover, hold most of the senior water licences in Alberta's "first-in-time, first-in-right" water allocation system. During May–June 2007,

Rocky View and the WID Board negotiated a deal in which Rocky View would obtain entitlement for just over 2.2 million cubic metres of water in exchange for $15 million, which the WID would use to convert fifty kilometres of its open canal into pipeline. The WID Board estimated that the amount of water saved from seepage and evaporation from the old canal would more than make up for the amount of water traded away, leaving it with a net water gain (Lacombe 2007; Alberta Environment 2010).

When the WID Board presented the deal to its membership on 21 June, it received a cool reception. Some irrigators worried about trading away part of their permanent water licence in exchange for short-term infrastructure improvement. Others worried that it might leave them water-short in drought years, with memories of the 2000–2 drought still fresh in their minds. Enough opposition surfaced that the deal was put to a membership plebiscite. The vote was held on 2 August, and the deal was approved by 57 per cent of the voting membership (D'Aliesio, "Part Two" 2007). Finally, after almost two years of searching, Rocky View and the CrossIron project were about to get the water they desperately needed.

The WID–Rocky View trade was approved by Alberta Environment in September, but, as a condition of the approval, 10 per cent of the volume of water traded (about 246,000 cubic metres) was held back and reallocated for environmental conservation (Alberta Environment 2010). Known as a conservation holdback, this was another novel feature introduced by the Water Act and the SSRB Water Management Plan. The Water Act empowered Alberta Environment to withhold up to 10 per cent of the volume of any water licence traded and to use this water to achieve water conservation objectives, the water conservation objectives themselves having been formalized in the SSRB Water Management Plan. Accordingly, the water held back from the WID–Rocky View trade was given its own licence and reallocated towards the achievement of the Bow sub-basin's water conservation objectives. Water that had been originally allocated for irrigation was now being redirected to commercial and environmental uses, a change emblematic of the wider socio-economic transformation in Alberta.

After withstanding a challenge at the Alberta Environmental Appeal Board in late 2007, the trade was completed and Rocky View finally had its water (Rocky View County 2007). CrossIron Mills, the massive shopping mall at the centre of the project, had its grand opening on 19 August 2009, two years behind schedule (CrossIron Mills 2009). The United

Horsemen's racetrack and casino has yet to be constructed, though an injection of cash from a casino investor in late 2012 promised to get it underway (Zickefoose 2012).

Though interesting in its own right, the CrossIron development story is recounted here to illustrate a new approach to water governance that has emerged in southern Alberta over the last twenty-five years. For nearly a century, water regulators dealt with water scarcity by continually seeking new water supplies to satisfy growing demands. This supply-side approach generally relied on "hard" infrastructure solutions, such as the construction of dams, canals, and pipelines, allowing more water to be controlled, stored, and diverted to where it was needed. Rocky View, for instance, saw nothing unusual about importing water for the CrossIron development from another sub-basin over two hundred kilometres away: they needed water and the Red Deer had surplus water, so why not bring in the supply to satisfy the demand? By the mid-1990s, this traditional "hard" approach to water governance began to give way to a softer approach that addressed water scarcity from the demand side of the equation. Rather than continually finding new water supplies – a task that had become increasingly difficult and costly – the "soft" approach focused on reducing or reconfiguring water demands to meet available supplies. The moratoria on new water licences and the market for water licences, where CrossIron eventually found its water, are two examples of this "soft" approach.

At the same time, Alberta has also pulled back in its use of hard water policy instruments. Since the completion of the Oldman Dam in 1992, only two moderately sized dams have been constructed in southern Alberta, one of which had been in planning and development since the late 1980s. Together, these dams total 106.5 million cubic metres of water storage capacity. This is in contrast to the two decades prior to 1992, a period in which 713.3 million cubic metres of water storage capacity was constructed. Dam construction has declined dramatically as the government has turned to softer policy instruments in its efforts to cope with perennial water scarcity.

While the shift from hard water policy to softer water policy has been significant, it is important not to overstate its extent. Alberta has not entirely abandoned the hard policy approach. There has not been, for instance, a widespread decommissioning of dams. Dams and other hard infrastructure remain key policy instruments of water governance in southern Alberta, and this seems unlikely to change any time soon.

Neither has Alberta fully embraced the soft policy approach. A true soft policy approach takes an ecosystem perspective to water governance and elevates sustainability as the central objective (Brooks and Brandes 2011). In pursuit of this objective, soft policy relies "on human ingenuity to find ways around current natural resource use patterns without losing the benefits of economic development that have improved the quality of life for so many people" (326). Though Alberta has certainly adopted innovative policy instruments designed to improve water use efficiency and conservation, one would be hard pressed to argue that sustainability has become the central water policy objective.

This is why Alberta's water policy reforms are characterized here as a change of degree to soft*er* policy rather than a change in kind to soft policy. Alberta's water policy now contains both hard and soft elements, but it is much softer than it was twenty-five years ago. The introduction of licensing moratoria, licence trading, conservation holdbacks, and water conservation objectives, in concert with a steep decline in new dam construction, constitutes a major change in water policy that, although not decisively soft, is certainly much softer than in the past.

The transition to a softer approach to water governance has been noted by many observers of Alberta water policy. Legal scholars, such as Percy (1996–7, 2005), have noted the fundamental changes in Alberta water policy brought about by the Water Act. This landmark legislation, he suggests, has provided regulators with important policy instruments for managing water demands, instruments that were adopted "only after all efforts at augmenting the natural supply of water had been exhausted" (2005, 2097). Economists such as Bjornlund, Nicol, and Klein (2007), Horbulyk (2007), and Nicol and Klein (2006) have also remarked on Alberta's significant water policy reforms, focusing mostly on the introduction of water licence trading in southern Alberta. Horbulyk, for instance, describes water licence trading as adding a needed "pressure-release valve" for water demands that could not be accommodated in the old, supply-oriented approach (Horbulyk 2007, 215). In short, there is general agreement that softer, demand-side water policies have become much more prevalent in southern Alberta since the 1990s, and that this represents a major policy change.

The purpose of this book is to explain how and why this policy shift occurred.

The widely presumed explanation is that Alberta simply ran out of hard policy options and shifted to softer policy options in the 1990s as a matter of necessity. This is what Percy suggests, for instance, in writing that Alberta had run out of options for augmenting its natural water supply. Yet, in actuality, Alberta still had a number of hard policy options at its disposal, but chose not to pursue them. One option was the Prairie Rivers Improvement, Management, and Evaluation (PRIME) project, an inter-basin diversion scheme that would bring water from the North Saskatchewan River Basin to the water-scarce south. Though PRIME is derided by critics as prohibitively costly and environmentally damaging, it resurfaces in policy discussions every time southern Alberta experiences a prolonged drought, most recently in 2000–2 (Struzik and Brooymans 2002). Moreover, comparable inter-basin diversion projects have been constructed in places like Australia and California, so the PRIME option, or something like it, cannot be dismissed out of hand. Alberta also had more dam building options at its disposal, such as the proposed Meridian Dam on the mainstem of the South Saskatchewan River, a project also considered during the 2000–2 drought (Canadian Broadcasting Corporation 2002). Clearly, the Alberta government had hard policy options at its disposal and it had shown a willingness to fund costly hard policy options right up to the mid-1990s, so a dearth of hard policy options is not a satisfactory explanation for the shift to softer water policy.

A more recent explanation for the water policy changes in southern Alberta has been proposed by Shapiro and Summers (2015), using a combination of path dependency and policy paradigms theory. Though path dependency helps to explain some of the longstanding stable features of southern Alberta water policy, path dependency, by definition, does little to account for policy change and cannot explain the significant policy reforms of recent years. Shapiro and Summers attempt to fill this gap by using Hall's policy paradigms theory, but it is not clear that the ideational changes they point to actually constitute proper policy paradigms. Accordingly, the shift to softer water policy in southern Alberta still requires explanation and this is the task taken up in this book, a task that takes us deep into Alberta's water politics and water policymaking.

Water Governance in Southern Alberta

Water scarcity is a fact of life in southern Alberta. Scarcity occurs when demands on a resource outstrip the available supply. In this way, scarcity has both a supply dimension and a demand dimension, and scarcity can be created (or exacerbated) when the supply of a resource decreases, the demands on a resource increase, or both happen simultaneously (Homer-Dixon 1999). Some places, such as southern Alberta, are more vulnerable than others to water scarcity simply as a result of their climate. Arid to semi-arid climates, where precipitation averages less than 500 millimetres per year, have both supply side limitations and demand side pressures that make them especially vulnerable to water scarcity. On the supply side, available water sources tend to be small because of the limited precipitation. On the demand side, pressures on these water sources are elevated because activities that would normally be rain-fed, such as agriculture, must rely on extractions from water sources instead. Thus, arid and semi-arid regions face a "double whammy" that makes them particularly vulnerable to water scarcity, and southern Alberta is well illustrative of this.

With average annual precipitation ranging from 300 to 500 millimetres, southern Alberta is one of the few regions in Canada with an arid to semi-arid climate (Matthews and Morrow Jr 1985, 38). It is important to note that this average masks a substantial degree of variability, with some years receiving much more or much less precipitation than the average. For instance, 2005 and 2013 were extremely wet years with extensive flooding along the Bow River and in the City of Calgary and its environs, while 2000, 2001, and 2015 were very dry years resulting in water shortages and depleted rivers throughout much of the region (Byrne, Kienzle, and Sauchyn 2010, 73). Given this variability, it is more accurate to describe water scarcity in southern Alberta as recurring and episodic rather than constant and continuous, and flooding has been just about as destructive and disruptive in the region as droughts have been. Nevertheless, limited precipitation is the norm in southern Alberta, making it vulnerable to the double whammy of water scarcity noted above.

On the supply side of this double whammy, southern Alberta's water sources are limited in their size and reliability.

There are two river basins in the region, the much larger South Saskatchewan River Basin (SSRB) and the much smaller Milk River Basin (see figure 1.1). The SSRB itself has four sub-basins, the Red Deer, the Bow, the Oldman, and the mainstem of the South Saskatchewan. The waters of the basin generally flow in an easterly direction, from the Rocky Mountains and its foothills, through Alberta and into Saskatchewan and Manitoba, before finally draining into Hudson's Bay. The Milk River Basin has only one substantial river and occupies a relatively small area along the province's southeastern border with Montana. The Milk

Figure 1.1 The River Basins of Southern Alberta

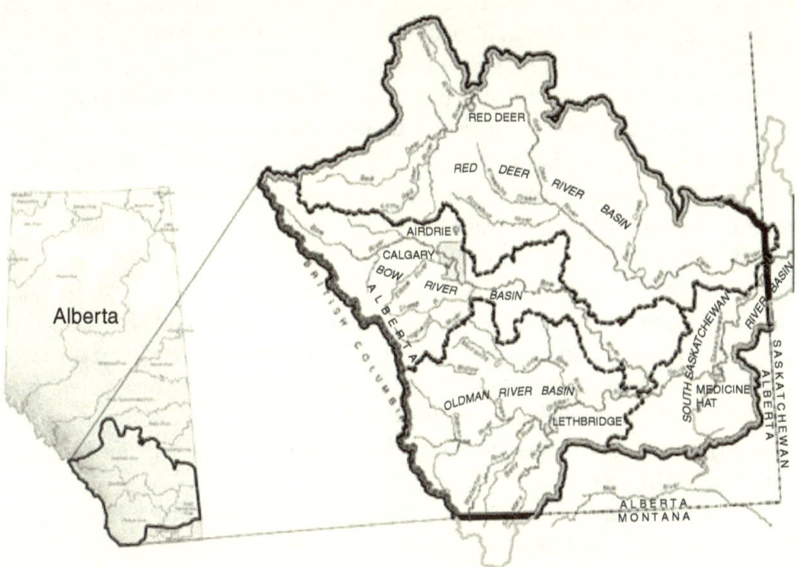

Source: Alberta Environment, *South Saskatchewan River Basin Water Information Portal* 2013.

Basin actually drains into the Mississippi system in the United States and has no natural hydrological connection to the SSRB, though the two basins are connected through a canal linking the St Mary and Milk Rivers in Montana. Both basins are shared by Alberta and Montana, adding a layer of international complexity to water governance in the region.

Much like the region's precipitation, river flows in the South Saskatchewan and Milk Basins are limited and variable, with flows typically highest during the spring freshet and lowest in late summer, but varying substantially from year to year and from river to river. This is why many Prairie residents wryly describe their rivers as either "mud or flood." The rivers of the SSRB have their source in the Rocky Mountains and are fed by a mixture of glacial melt, snow melt, and rainfall run-off. The Milk River, in contrast, has no glacial source and relies almost entirely on run-off. Accordingly, flows tend to be more variable in the Milk than in the South Saskatchewan, though both basins are vulnerable to prolonged low flow periods.

Climate change is a wild card in the region that could reduce natural water supplies even further. Most climate change models predict that the region's climate will become substantially warmer – exceeding the global average increase in temperature – and that precipitation patterns will shift so that winter months will have more rainfall and less snowfall and summer months will have less precipitation altogether (Byrne, Kienzle, and Sauchyn 2010, 64–5). Less snowfall and more rainfall in the winter months promises to increase winter flows and reduce spring/summer flows, the time of year when water demands are highest (Barnett, Adam, and Lettenmair 2005). The warming climate also threatens to melt the Rocky Mountain glaciers that are a small but reliable water source for the SSRB. These glaciers have been receding for a number of years and are projected to disappear altogether in a matter of decades. Overall, the median climate change scenario predicts that, by mid-century, mean annual flow in the South Saskatchewan River at Medicine Hat could be reduced by 8.5 per cent (Byrne, Kienzle, and Sauchyn 2010, 68).

While the supply of water in southern Alberta is limited and seems likely to shrink further, the demands on these water sources are limitless and seem likely to grow.

For over a century, the biggest demand on southern Alberta's water has come from irrigated agriculture. Irrigation currently accounts for about 77 per cent of water use in the SSRB, the next closest water use being municipal water usage at 14 per cent (Alberta Economic Development Authority 2008, 14). Much of the region's irrigation is concentrated in its thirteen irrigation districts, which account for over 80 per cent of irrigation in the SSRB (Alberta Agriculture 2010). The irrigation districts are farmer-run cooperatives with extensive and sophisticated water diversion, storage, and transportation systems that bring water to irrigators, often from considerable distances. The remaining 20 per cent of irrigation is undertaken by private irrigators outside of the districts, who gain access to irrigation water through their own pumps or weirs. The vast majority of the crops grown using irrigation are field crops (or "cash crops") such as wheat, canola, and soybeans; a smaller proportion is dedicated to growing hay and other fodder crops, and a minute proportion is used to grow vegetables, such as the famous Taber sweet corn (Babooram 2011). Irrigation is a particularly heavy water demand because it requires large volumes of water, it consumes most of the water that it uses, and its water demands are usually highest during dry periods, precisely when river flows are at their lowest (Postel 1999).

As the population and economy of southern Alberta has grown by leaps and bounds over the past century, other economic water demands have also emerged. One of these is the use of water for oil extraction; however, many are surprised to learn that the oil industry is a very small water user in southern Alberta. Conventional oil extraction in southern Alberta is not nearly as water intensive as unconventional oil extraction in the northern oil sands, and there are far fewer oil-related water demands in the south as compared to the north. A much bigger water demand comes from the growing urban and suburban areas. The population of Calgary and its environs has nearly tripled since the early 1970s, and growing cities need water for domestic and sanitary purposes, to support industries and commerce (such as the CrossIron project), and even to maintain recreational facilities such as arenas, swimming pools, and golf courses. They also need electricity, some of which is generated by hydroelectric dams in the headwaters of the Bow sub-basin. As a group, these newer economic water demands use much less water and produce higher economic returns per water unit than irrigated agriculture. This has created some pressure to reallocate water from lower-value water users (irrigators) to higher-value water users (cities and industries), adding an element of urban-rural rivalry to the water demands in southern Alberta.

There are also a growing number of water users who value water that is left instream, for a variety of reasons. Some of these water users (such as tourism operators) depend on instream water for their livelihoods, while others (such as anglers, hunters, boaters, and hikers) rely on instream water for recreational purposes. Environmentalists also highly value instream water but for non-instrumental reasons. They argue that riverine and riparian environments have inherent value as part of the region's natural heritage, and that the water needs of these environments should take priority over all other water uses. Though these environmental needs have always been there, the notion that they should be recognized and prioritized is a relatively recent one. Conservationists, such as the Alberta Wilderness Association and the Federation of Alberta Naturalists, have worked tirelessly since the 1970s to represent these beliefs in Alberta's water politics. In this way, the environment and its water needs is something of a new water demand in southern Alberta, a demand that directly confronts the economic water uses – particularly the large, consumptive irrigation uses – and is rivalrous with them.

A final group of water demands comes from southern Alberta's First Nations. The First Nations use water for a variety of personal, commercial, and agricultural purposes, but they also have close spiritual and cultural attachments to water in its natural state. Having occupied the area for centuries prior to European settlement, First Nations recognized water as a lifeblood in the semi-arid climate and had particular reverence for it, an attachment that carries forward to today. A lot of uncertainty still prevails around First Nations water rights claims in the region. Some claims have been settled, such as the 2010 agreement between the governments of Canada, Alberta, and the Piikani First Nation (formerly known as the Peigan First Nation), which recognized the Piikani's rights to water from the Oldman River for residential and agricultural needs and transferred over $200 million to the band in restitution for land and water wrongly taken from them. In exchange, the Piikani agreed to obtain water licences from the Alberta government for all of its commercial water uses and to give up any future water rights claims (Phare 2009, 1–6). Other First Nations, such as the Stoney Nakoda and the Tsuu T'ina and Samson Cree, still have ongoing water rights disputes that remain unresolved (Laidlaw and Passelac-Ross 2010). Although the First Nations are small water users in southern Alberta, most of their water rights claims are high priority – claiming to have precedence over all other uses – and thus have the potential to upset the delicate balance among southern Alberta's other water demands.

Fears of water demands outstripping water supplies date back almost a century in southern Alberta. Water scarcity fears have usually found expression as concerns about over-allocation, the allocation of more water (through government licences) than is physically available in a given stream. Over-allocation problems were first detected in the summer of 1919 when federal officials warned that, on some rivers, so much water had been allocated for irrigation that new communities would not be able to secure water supplies for essential domestic purposes (Percy 1996–7, 225). As irrigation continued to grow, over-allocation fears intensified on some rivers and spread to others. The worst affected areas were in the southwest of the province on the southern tributaries of the Oldman River. By the late 1970s, Alberta Environment declared rivers in this area closed to new water licence applications simply because there was no water left to allocate. The Waterton River was allocated at 75 per cent of its median annual flow, the Belly River was allocated at 80 per cent of median annual flow, and the St Mary River was allocated at a whopping 118 per cent of median annual flow. Other heavily used rivers were marginally better, the Oldman River at 70 per cent of median annual flow and the Bow River at 68 per cent (Alberta Environment, "South Saskatchewan River Basin Water Allocation" 2003, 12).

Clearly, water scarcity is an omnipresent challenge in southern Alberta, crucial to the region's long-term economic prosperity and environmental sustainability; it also creates a "wicked problem" for those involved in water governance. To label water governance in southern Alberta as a wicked problem is not to pass judgment on it or the people involved. Instead, a wicked problem is a type of governance problem "characterized by a high degree of scientific uncertainty and deep disagreement on values" (Balint et al. 2011, 2). These two factors, in combination, make wicked problems highly intractable for the actors seeking to manage them, hence their "wickedness."

Because of the high degree of scientific uncertainty, there are no readily identifiable or universally acceptable "technical" solutions to wicked problems. Without scientific signposts to guide them, actors tend to view wicked problems through normative lenses based on their respective values and beliefs. However, there is also deep disagreement on values, meaning that actors view the same problem through different world views, resulting in wildly varying problem definitions and little agreement on acceptable problem solutions. Ultimately, the actors involved in wicked problems usually retreat to various normative camps, each advocating its own definition of the problem and its preferred solutions, often talking past each other, and finding it difficult to reach common ground.

Water governance in southern Alberta has all of the hallmarks of a wicked problem. Despite some claims to the contrary, the application of science cannot determine how much water should be extracted from southern Alberta's rivers for human use and how much should be left to preserve riverine environments. There is no scientifically "optimal" or "technical" solution to this basic governance problem. In the face of this scientific uncertainty, the matter becomes a fundamentally normative one, heavily influenced by actors' world views. The differences between the hard and soft approaches to water governance are illustrative of this. The hard approach advocated by irrigators and engineers views water as a resource to be exploited and controlled by humans, useful only insofar as it can be used to support economic development. In this view, the basic problem is supplying water to human users, and most of the solutions involve the construction of hard infrastructure to provide this supply. The soft approach advocated by environmentalists and environmental economists, in contrast, puts much greater value on water in its natural state, with the problem defined as riverine environmental protection, and the solutions involving a curtailment in human water demands. Both groups are looking at the same rivers – and sometimes looking at the same river data – but seeing different problems and pointing to conflicting solutions. This is the essence of a wicked problem.

The governance of wicked problems, then, typically becomes a contest of competing policy visions, each with a different normative basis, and none of which are inherently or technically superior to the others. Each camp may try to claim technical superiority, and they may genuinely believe in their superiority, but these claims are more rhetoric than reality. There is no scientific or technical basis, for example, for choosing between the irrigators' vision of economic prosperity and the environmentalists' vision of environmental preservation. Governmental decision-makers, as sovereigns, are left to choose among the competing normative policy visions or to somehow try to reconcile them. Governmental policy decisions are crucial because these decisions elevate some visions over others and put governmental resources behind them, through budgetary allocations, program creation, and rule development and enforcement. The stakeholders advocating various policy visions recognize this and dedicate a lot of time, effort, and money to try to influence the policy choices of governmental decision-makers. In this way, the management of wicked problems is both inherently and intensely political.

The shift from hard to softer water policy suggests that something significant has changed in Alberta's water politics over the past few decades. It even further implies that this shift has something to do with the role and influence of environmentalists, the main proponents of soft water policy in southern Alberta. To identify and understand how the influence of environmentalists has increased and how the shift to softer water policy was achieved, this study employs the Advocacy Coalition Framework (ACF) as an explanatory theoretical lens.

The ACF is a theoretical framework for understanding policy development that has existed since the late 1980s and has its origins in the work of two American political scientists, Paul Sabatier and Hank Jenkins-Smith. Sabatier and Jenkins-Smith developed the ACF specifically to understand how policies are made in wicked problem situations (Sabatier and Weible, "Advocacy Coalition Framework" 2007, 189). This makes the ACF particularly well-suited for studying water policy development in southern Alberta. The framework discards any notion that the policy process is a technocratic search for an optimal policy solution, and instead focuses on actors' beliefs and their abilities to translate these beliefs into policy by influencing government decision-makers. The ACF purports that actors sharing similar beliefs work together in advocacy coalitions, and it is these coalitions that work, through various means, to shape government policy decisions. Policy changes, such as the shift to softer water policy in Alberta, are broadly explained as resulting from changing coalition beliefs and/or changes in the coalition balance of power. However, the ACF is a complex and multifaceted theoretical approach with multiple avenues of ongoing research that defy quick summary and description. For this reason, chapter 2 is dedicated to outlining the main tenets of the ACF as well as the general and specific ACF-based hypotheses examined in this study.

Methodology and Book Outline

This book examines the shift to softer water policy in southern Alberta through a diachronic case study using primarily qualitative methods of data collection and analysis. Following Gerring's definition, a diachronic case study involves the study of a single case over a defined period of time (Gerring 2004). If properly designed, a diachronic case study allows for variance of key variables over time so that within-case comparisons can be made between defined time periods. This is the most common case study approach used by social scientists and is more popularly known as a pre/post comparison (Gerring 2007). In this book, within-case comparisons are made between the hard policy era and the softer policy era in an effort to determine what political, social, economic, and environmental factors shifted over time and contributed to the onset of softer water policy.

The book's period of study spans from 1971 to 2007, beginning with the election of the first Progressive Conservative (PC) government in Alberta and ending with the implementation of the SSRB Water Management Plan. This period effectively captures the shift from hard to softer water policy. It also sets up within-case comparisons in which the development of hard policies (such as the construction of the Oldman Dam and the creation of the SSRB Water Management Policy) can be contrasted with the development of softer policies (such as the passage of the Water Act and the negotiation of the SSRB Water Management Plan).

The processes can then be compared to determine what factors shifted over time and contributed to the onset of softer water policy.

The defined period of study also allows for some key variables in policy development to be held relatively constant. The PCs were first elected to government in Alberta in 1971 and remained in power continuously until 2015, so the period of analysis is defined by a period of PC government. Furthermore, shortly after the PCs first came to power, Ottawa negotiated a retreat from irrigation assistance in southern Alberta, leaving the province as the only water policymaking venue during most of the period of study. In this way, the period of study is characterized by provincial water policymaking by PC governments, holding some important policy process variables (relatively) constant and allowing the policy impacts of advocacy coalitions to be more clearly isolated.

The book relies primarily on qualitative methods of data collection and analysis, using a combination of process tracing and content analysis. More specifically, the overall approach is one of process tracing, with content analysis relied upon to investigate and measure some of the central variables in the process tracing analysis.

Process tracing is an analytical method used to reconstruct and identify causal processes, in this case the causal process resulting in the shift to softer water policy. The aim is to get inside a causal relationship and gain intimate knowledge of how and why independent and dependent variables are causally related, often through multiple intervening variables (George and Bennett 2005). When the independent variables are largely unknown, process tracing is more exploratory and focuses on identifying causal variables and how they contribute to the dependent variable of interest. Gerring describes process tracing as akin to "detective work" in which "the analyst seeks to make sense of a congeries of disparate evidence, each of which sheds light on a single outcome" (Gerring 2007, 178). The process tracing in this book is theoretically grounded in the sense that it is conducted through an ACF lens, and particular emphasis is placed on the empirical presence and causal impact of ACF-related variables such as advocacy coalitions and coalition resources.

Given that it is used to reconstruct past causal processes, process tracing has a historical bent and usually relies on disparate sources of historical evidence to gain as complete a picture as possible of the processes of interest. In this study, a variety of historical data were used, including government reports, the *Alberta Hansard*, contemporary participant accounts, contemporary newspaper articles, surveys, and secondary analyses by other social scientists. These archival data were supplemented with confidential, semi-structured interviews with a variety of participants in the Alberta water policy subsystem, including members of both advocacy coalitions as well as members of governmental and non-governmental organizations. All of these data were compiled into timelines documenting the development of important water policies, both hard and soft. Following the tenets of good process tracing, no piece of data was entered into the timelines until it was triangulated, that is verified by three independent sources (Bennett 2008, 707). This provided confidence in the validity of the timelines, and the timelines provided an increasingly fulsome picture of the factors at work in the development of southern Alberta water policy.

Within the process tracing analysis were some key ACF variables – policy-relevant beliefs and advocacy coalitions – that could not be adequately evaluated using process tracing alone. Instead, these variables were investigated using qualitative content analysis, and the results fed into the larger process tracing analysis. The results of the content analysis are presented in chapter 4 and involve a systematic analysis of stakeholder submissions during policy consultation processes in 1978, 1995, and 2002/5. This analysis shows the prevalent policy-relevant beliefs at each point in time and provides a basis for identifying groups of like-minded actors and, ultimately, advocacy coalitions. With this solid empirical grounding, these advocacy coalitions (which I have labelled the Aggies and the Greens) become the main protagonists in the process tracing analysis presented in the rest of the volume. An overview of the content analysis is provided in chapter 4, but a more detailed account is provided in the book's appendix for those interested in this part of the study's methodology.

After exploring the ACF and establishing the study's hypotheses in chapter 2 and describing the shift from hard to softer water policy in more detail in chapter 3, the remainder of the book can be roughly divided into two sections. The first section includes chapters 4 and 5, which examine key ACF variables and establish their empirical validity. As mentioned above, chapter 4 examines the prevailing policy-relevant beliefs and advocacy coalitions at three different points in time (1978, 1995, and 2002/5), using this within-case comparison to gain a clear sense of what advocacy coalitions existed, how these coalitions may have changed over the period of study, and how this shift in advocacy coalitions contributed to the shift to softer policy. Chapter 5 takes up the concept of advocacy coalition power, examining the various coalition resources identified in the ACF and how the coalition balance of power changed over the period of study. The second section of the book, which includes chapters 6, 7, and 8, then puts these key ACF variables into action by focusing on specific water policy processes. Chapter 6 examines water policy development during the last two decades of the hard policy era, chapter 7 examines the initial shift to softer water policy through the passage of the Water Act, and chapter 8 focuses on the negotiation of the SSRB Water Management Plan. Chapter 9 closes the book by examining some of the impacts of the shift to softer water policy for water users in southern Alberta, by revisiting the study's hypotheses, and by exploring the implications of the study's findings for the ongoing development of the ACF and the governance of wicked problems more generally.

Chapter Two

The Advocacy Coalition Framework

Scholars are interested in understanding the world. In order to accomplish that, they must use conceptual frameworks to help tell them what is probably important and what can be ignored.

Paul Sabatier and Hank Jenkins-Smith,
Policy Change and Learning: An Advocacy Coalition Approach

To describe policy studies as theoretically crowded is something of an understatement. The third edition of *Theories of the Policy Process*, the preeminent student text in the field, includes chapters on no fewer than eight widely followed models, theories, and frameworks of policy development. Moreover, it does not include a number of potentially viable but less widely followed approaches, and it purposefully excludes all of the critical approaches, which are well beyond its positivist scope (Sabatier and Weible, *Theories* 2007). Compared to microeconomics, where the rational choice approach is still firmly ensconced and challenged only at the margins by the behavioural and critical approaches, the field of policy studies seems practically chaotic. There is little chance that some "master theory" of the policy process will emerge anytime soon, nor is it clear that one is even desirable (Cairney 2013).

Selecting from the menu of theoretical options is a challenge for policy scholars. Often this choice is based on little more than familiarity. Young policy scholars are trained in the tenets and methods of the approach used by their mentors, and they stick with this approach because they are invested in it and the costs of switching are quite high. For those who ignore the switching costs and try to make a more reasoned decision, the choice ultimately comes down to how one conceives of the policy development process and what assumptions one is most comfortable making. All theories necessarily make assumptions and these assumptions simplify and caricaturize policy development in order to make it explicable. The assumptions usually create some theoretical blind spots – aspects of policy development that a theory overlooks or cannot account for – and one has to be comfortable with a theory's blind spots in order to work with it. It is the existence of these blind spots, and theorists' attempts to remedy them, that explains the multitude of theories that exists in policy studies.

In my evaluation, the ACF is an appealing theoretical approach because it has few major blind spots. In his seminal work, Hugh Heclo describes two faces of policy development, "powering" and "puzzling." The "powering" face conceives of policy development as an exercise of raw political power, with contending individuals and groups seeking to out-compete each other in capturing policy. The "puzzling" face conceives the policy process as an exercise in collective problem-solving, trying to redress the major social ills of the day (Heclo 1974). Both faces capture an essential aspect of policymaking, and ignoring either of them, in my estimation, creates a major theoretical blind spot. The ACF seems to be one of the few theoretical approaches capable of simultaneously recognizing and embracing both the "powering" and "puzzling" faces of policy development, even though it is still working out their intersection and interplay. For this fundamental reason, I have grown increasingly comfortable with the ACF and have chosen it as the analytical framework for this study.

This chapter describes the main concepts and the central tenets of the ACF, followed by an exploration of the major lines of ongoing theoretical development within the framework. The aims of this chapter are to provide the reader with some familiarity with the ACF and, ultimately, to specify the ACF-based hypotheses that will be investigated in the remainder of this book to explain the shift to softer water policy in southern Alberta.

An Introduction to the ACF

As briefly described in chapter 1, the ACF is a framework for conceptualizing, understanding, and theorizing about public policy development. Though it was originally developed to explain policy development in the American context, the founders and disciples of the ACF have shown a great willingness to apply the framework beyond its originating jurisdiction, and there are now ACF applications in countries around the world, even in some non-democratic countries. Adherents of the ACF have also shown a willingness – a willingness that is somewhat unusual compared to other theories of the policy process – to revise the framework in light of empirical evidence from its many applications. The framework has gone through at least two major revisions, done in an effort to make the theory more portable and its hypotheses more generalizable. It is the most recent version of the ACF that is used in this book and described in this chapter.

Although it originated in the country next door, the ACF has been applied only sparingly to cases of Canadian policy development. A review in 2009 found only six applications of the ACF to Canadian cases in peer-reviewed journals and books during the first twenty years of the ACF's existence (Weible, Sabatier, and McQueen 2009, 127). More have been published since this review, including Montpetit's (2011) cross-national study of biotechnology policy, Bratt's (2012) study of nuclear power policy in four Canadian provinces, Heinmiller's (2013) study of the development of the Alberta Water Act, and Swigger and Heinmiller's (2014) study of mental health policy in Ontario. But the overall number of ACF studies remains low, relative to the number of Canadian public policy scholars. The reasons for this are unclear but should not be attributed to any inherent flaw in the framework that makes it inapplicable to parliamentary contexts. In fact, the ACF has been successfully applied to explain policy development in European, Australian, and Asian parliamentary systems, and some of the revisions made to the framework were adopted to make the ACF more readily applicable to parliamentary policymaking (Sabatier and Weible, "Advocacy Coalition Framework" 2007). The relatively small number of Canadian applications of the ACF should be regarded as a research frontier, and this book makes a modest contribution in that regard.

The primary unit of analysis in the ACF is the policy subsystem. The ACF assumes the central importance of policy subsystems based on the sheer complexity of modern policymaking, particularly in wicked problems, and the tendency of actors to specialize and concentrate their efforts in one policy area (or a few). The ACF further assumes that most policy development occurs within policy subsystems, so understanding the political dynamics of policy subsystems is essential to explaining policy change. Within each policy subsystem are the various politicians, public servants, interest groups, journalists, scientists, and private citizens who claim an interest or a stake in the policy area and who regularly participate in policy processes. Policy subsystems also often include actors from multiple levels of government, particularly in federal states where there is substantial overlap of jurisdictional responsibilities between national and sub-national governments.

Defining the boundaries of a policy subsystem is an important empirical task for ACF analysts and can be difficult because subsystems overlap and nest. The guiding principle in defining subsystems is to "focus on the substantive and geographic scope of the institutions that structure interaction" (Sabatier and Weible, "Advocacy Coalition Framework" 2007, 193). In our case, the substantive boundaries are those pertaining to water allocation, and the territorial boundaries are defined by Alberta's portion of the SSRB. This southern Alberta water policy subsystem has existed for decades and is distinct from the water policy subsystem in the northern part of the province, which deals with very different challenges (such as pollution and loss of habitat from oil sands mining) and involves a very different cast of actors (oil companies rather than irrigators). The northern Alberta water policy subsystem is also much newer, essentially emerging with the development of the oil sands. The southern and northern water policy subsystems clearly have some overlap in actors and legislation, but they are territorially and functionally distinct. The focus in this book is on the southern subsystem, though, for the sake of expediency, it is often referred to as simply the Alberta water policy subsystem.

In the ACF, actors in policy subsystems are driven fundamentally by their beliefs and their desire to see their beliefs reflected in policy. The ACF assumes that actors' beliefs can be understood as operating at three different levels of abstraction. At the deepest level are deep core beliefs, which are quite broad in scope, predominantly normative, and very slow to change in response to empirical contradiction. These beliefs are relevant across a broad range of policy subsystems and include such things as one's left-right ideological orientation and one's attitude towards nature. At the next deepest level are policy core beliefs, which are intermediate in scope, because they involve the application of deep core beliefs to a specific policy subsystem. Irrigators and environmentalists, for instance, have different deep core beliefs about the value and uses of nature, and when these beliefs are applied in the Alberta water policy subsystem, they form the basis for their very different water policy preferences. Policy core beliefs tend to be resistant to change "but are more likely to adjust in response to verification and refutation from new experience and information than deep core beliefs" (Weible,

Sabatier, and McQueen 2009, 123). At the shallowest level are secondary beliefs, which are narrowest in scope and pertain to the details of various policy issues, such as preferred modes of policy implementation. Although secondary beliefs are linked to policy core beliefs, they are more empirically based and are therefore more amenable to change (Sabatier and Weible, "Advocacy Coalition Framework" 2007, 194–6).

Understanding actors' beliefs is important, because these beliefs form the basis of advocacy coalitions, one of the central explanatory variables in the ACF. Like-minded actors in policy subsystems very often make common cause in pursuit of their policy preferences simply because they are more effective working together than working in isolation. When a group of actors shares a set of policy core beliefs and engages in a non-trivial degree of collective action, they form an advocacy coalition (Sabatier and Weible, "Advocacy Coalition Framework" 2007, 196). The ACF hypothesizes – and research has largely confirmed – that advocacy coalitions tend to be relatively stable over time, despite turnovers in membership and periodic defections, so that, on most major policy controversies in a policy subsystem, there is a characteristic line-up of coalitions seeking to influence policy (Jenkins-Smith et al. 2014, 195).

The number of advocacy coalitions in a policy subsystem typically ranges between one and five and, when multiple coalitions are present, they compete with each other to influence government decision-makers and have their policy core beliefs reflected in government policy (Sabatier 1993). For instance, the Alberta water policy subsystem was, for decades, dominated by a single "Aggie" coalition, which believed that irrigation and agricultural development should be the first priority of water policy. Since the mid- to late 1980s, however, the Aggies have been joined by the "Greens," a new advocacy coalition, which believes that ecosystem protection should be the first priority of water policy. Since the Greens' emergence, the two coalitions have competed vigorously to see their respective policy core beliefs reflected in Alberta's water policies.

In most policy subsystems, there is a dominant advocacy coalition that holds the most sway over policy decision-makers, and one or more minority advocacy coalitions that have less power and often struggle to influence policy. Identifying the dominant coalition and its policy-oriented beliefs is a crucial task in ACF research, because the beliefs of the dominant coalition usually have a strong influence on policy design. As we will see in chapters 4 and 5, the Aggies have long been the dominant coalition in the water policy subsystem, and remain so. Accordingly, they have had the most influence over policy decision-makers, and this explains why many remnants of hard water policy still prevail in southern Alberta. Since their emergence, the Greens have been the minority coalition but have still managed to exert some policy influence, and their presence has been a very important factor in the shift to softer water policy.

In addition to advocacy coalitions, some policy subsystems also feature one or more policy brokers. Policy brokers are actors who position themselves between competing advocacy coalitions, often attempting to bridge the differences between them. Policy brokers come in many different forms, but they usually have some credibility with all of the contending advocacy coalitions and, as a result, can speak earnestly to all of them and serve as a conduit – and sometimes even a translator – between them. This can facilitate both policy-oriented learning and policy compromises between the coalitions. From the mid-1990s onward, the Alberta Department of the Environment took on the policy broker role between the Aggies and Greens. Since it housed both dam builders and environmental regulators, it had credibility with both advocacy coalitions and was instrumental in facilitating policy-oriented learning and forging policy compromises between them. In this way, Alberta Environment was an important actor in the shift towards softer water policy, even though it was not a member of either advocacy coalition.

As advocacy coalitions strive to influence policy decision-makers, they may utilize a number of different resources towards this end. Largely on the basis of work by Sewell (2005) and Weible (2005), Sabatier and Weible ("Advocacy Coalition Framework" 2007, 201–4) have proposed a typology of six main coalition resources that will be used here:

1. **Formal Legal Authority to Make Policy Decisions**. All advocacy coalitions ultimately work to have their beliefs reflected in policy, often by seeking to influence and persuade policy decision-makers. In some cases, however, a coalition may be fortunate enough to have some policy decision-makers in its membership. These actors can help coalitions secure veto points in the policy process, and they can advocate for a coalition's policy position in closed decision-making bodies such as executive cabinets, legislative committees, or intergovernmental forums. Examples of coalition actors who may have at least some policy decision-making authority include senior administrators, legislators, unelected executives, elected executives, and, in some cases, judges.

Formal legal authority to make policy decisions is the most powerful of all coalition resources, for reasons explained further below.

2. **Public Opinion**. In democratic contexts, public opinion can be a very important coalition resource. Most policy decision-makers are either directly or indirectly accountable to the public, and decades of scholarship has shown that these actors are quite cognizant of this accountability and that their behaviour is influenced by it. Accordingly, decision-makers are loath to make policy choices that are at odds with public opinion, and a coalition that can cultivate and draw on favourable public opinion has an important source of leverage over policy decision-makers.

3. **Information**. In most modern policymaking, information about the nature of policy problems, their severity, and the most appropriate policy solutions is itself contested and used strategically by policy elites. This is evident in the efforts of policy elites to control the release of information, to massage data, and to spin social and political events so that they bolster their policy positions. Information can be used by advocacy coalitions to rally supporters, to shape public opinion, or to sway policy decision-makers. One example is the importance of scientific information: when an advocacy coalition appears to have science on its side, the credibility of its policy positions is greatly increased.

4. **Mobilizable Troops**. Advocacy coalitions can sometimes find strength in numbers through the number of activists it can mobilize in support of its cause. These mobilizable troops are different from general public opinion, as the public may passively support a coalition, but mobilizable troops actively work on its behalf. These troops may engage in public demonstrations, work on election campaigns, help with fundraising drives, or participate in public education efforts, all of which can bolster a coalition's advocacy efforts.

5. **Financial Resources**. Money cannot buy policy, but it can buy coalitions a lot of resources towards this end. Coalitions with ample financial resources can purchase the services of professional lobbyists and public relations experts, they can support research and development, they can undertake extensive public education campaigns, they can employ full-time support and administrative staff, and they can donate to election campaigns. Clearly, money is a very important resource to any advocacy coalition, and those without it fight an uphill battle against those who have it.

6. **Skilful Leadership**. Charismatic individual leaders can be an important resource for any advocacy coalition. Skilful leaders can have a major influence on a coalition's internal cohesion, its capacity to mobilize

troops, its ability to sway public opinion, and its ability to raise money. The literature on policy entrepreneurs also shows how skilful and well-connected individuals are important for exploiting political opportunities and moving issues onto the policy agenda. Though the extensive literature on leadership provides little clear basis for defining or measuring leaders, there is no denying their intangible importance to advocacy coalitions.

Using this typology of coalition resources, Nohrstedt (2011) examined Swedish signals intelligence policy and drew some important conclusions about their relative importance. He found that some coalition resources were much more potent than others in allowing the advocacy coalitions to shape policy outcomes and suggested some adjustments to the ACF based on this finding. "One theoretical implication from this analysis is thus the need to introduce hierarchy in the description of political resources; some political resources are clearly more important than others as a basis for coalitions to achieve influence in the policy process. When seeking to explain why public policy programs change, formal legal authority should enjoy unique status compared with other political resources because legislators are veto players whose agreement is needed for policy change to happen" (480).

Building on this finding, Heinmiller (2013) subsequently argued that a more systematic means of conceptualizing the key distinction between the coalition resources is to incorporate Nye's concepts of "hard" and "soft" power into the ACF. These concepts succinctly capture the important distinction noted by Nohrstedt while providing a conceptual foundation for measuring coalition power and developing propositions about it.

Hard power is command-based or coercive power, the ability to push others to do things they might not do of their own volition (Nye 2011). In this sense, hard power is based fundamentally on the use of force or the threat of the use of force, which means different things in different contexts. In the international system, the context analysed by Nye, there is no global sovereign monopolizing the legitimate use of force. So the actors in the international system (mostly states) are unconstrained in their use of force on each other, and the ultimate hard power resource is a strong military. In domestic and democratic policy subsystems, the context is very different. Policy subsystems are nested within a sovereign state that monopolizes the legitimate use of force, and the actors in a subsystem (mostly policy elites) do not inflict or threaten to inflict violence on each other. In this context, the ultimate hard power resource is the formal legal authority to make policy decisions, since this authority is implicitly backed up by the state and its monopolization of the legitimate use of force. It is also this feature that makes formal legal authority unique and the most potent resource that an advocacy coalition can possess.

Soft power is more co-optive than coercive and is based on the ability to pull or attract people towards a desired action (Nye 2011). Actors using soft power rely heavily on their credibility, legitimacy, and persuasive abilities. The exercise of soft power is not substantially different in the international and domestic contexts, as the tools of positive persuasion are fairly universal. Soft power resources tend to be more diverse and ephemeral than hard power resources and, in a domestic policy subsystem, would include things like public opinion, information, mobilizable troops, and skilful leadership. Any or all of these resources can attract or cajole policy decision-makers towards a particular policy, but none of them have the authoritative force of hard power.

Financial resources are a special case, as they can be transformed into either hard power or soft power resources, depending on how they are used. In domestic policy subsystems, however, money is much more readily converted into soft power than hard power. Whereas states in the international system can easily convert their money into hard power through military purchases, policy elites in advocacy coalitions do not have this option. Policy elites may be able to use their financial resources to gain access to policymakers and may use campaign donations to gain some leverage over them, but, short of outright bribery, they cannot purchase the formal legal authority to make policy decisions. Accordingly, it is quite difficult to convert financial resources into hard power in domestic policy subsystems, and, most often, advocacy coalitions convert money into soft power resources instead.

In terms of sheer ability to shape policy outcomes, hard power is the more potent force, because it allows coalitions to push through (or protect) their policy goals by virtue of their influence over key decision-making points.[1] In this sense, the dominant advocacy coalition in any policy subsystem is the one that holds the preponderance of hard power resources. Nevertheless, as Nye has repeatedly argued, the

1 In the case considered here, the key decision-making points are the legislative and executive branches in Alberta. In other cases, the courts may also be a key decision-making point, and influence here might also be considered a form of hard power.

influence of soft power should not be underestimated. Minority coalitions with soft power resources may use them to undercut the legitimacy of opposing coalitions, to attract new members to their fold, and to persuade opponents of the inherent "rightness" of their policy positions. The use of soft power may be a more indirect and longer-term means of influencing policy than the use of hard power, but it can be quite effective. And, for a minority coalition with little hoping of displacing a dominant coalition, it may be their only means of influencing policy. This has been the case in the Alberta water policy subsystem where the Greens, as a minority coalition, have relied almost exclusively on soft power to get their environmental protection and restoration objectives enshrined in water policy.

As figure 2.1 shows, policy subsystems are the primary unit of analysis, but the ACF recognizes that policy subsystems are significantly affected by forces well beyond their borders. Within the framework, these forces are categorized as relatively stable parameters, external events, and long-term coalition opportunity structures.

The relatively stable parameters refer to features in the natural, social, and legal context that are relatively constant, such as the distribution of natural resources, the prevailing sociocultural values, and a state's constitutional structure. Significant changes in any of these factors – which are relatively rare – can have a major impact on policy subsystems and the advocacy coalitions within them. For instance, over the last fifty years, Alberta has experienced massive (oil-based) economic growth, extensive urbanization, and a societal shift to post-materialist values. As shown in chapters 4 and 5, these "big and slow" changes (Pierson 2004) gradually but significantly realigned the advocacy coalitions and coalition resources in the water policy subsystem, resulting in the emergence of the Greens as a political force in water policy development.

External events refer to factors outside of policy subsystems that are more changeable than the relatively stable parameters, but can still have a significant impact on a policy subsystem. These events include changes in socio-economic conditions (such as the onset of a recession); changes in public opinion (such as periodic swings in public support for environmental causes); changes in government (after an election); and policy decisions and impacts from other policy subsystems. In this way, the ACF situates policy subsystems within their ecological, social, and political contexts and is cognizant of how changes in the world outside of policy subsystems can significantly affect policy development. Yet the impact of these external factors is indirect, as they are

Figure 2.1 The Advocacy Coalition Framework

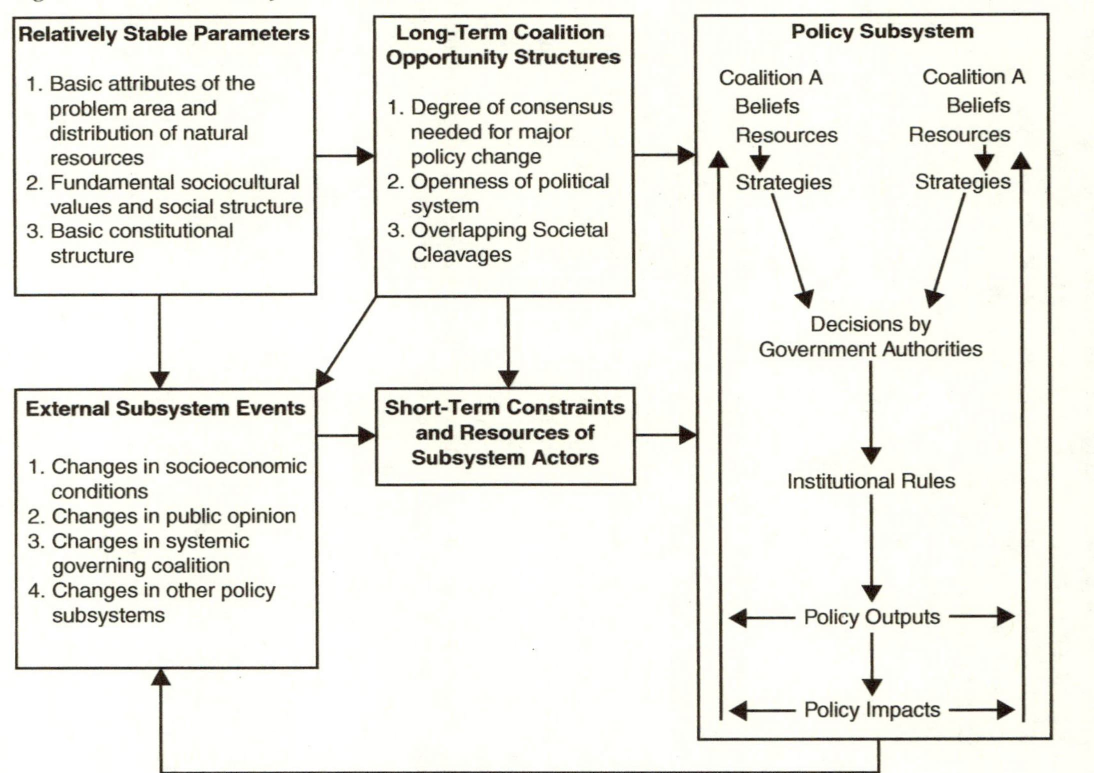

Source: Weible et al. 2011, 352.

refracted through policy subsystems (Sabatier and Weible, "Advocacy Coalition Framework" 2007, 193). There are many examples of external events shaping water policy development in southern Alberta, as will be shown in chapters 6, 7, and 8. One illustrative example is the drought in 2000–2, which added sudden urgency to water policy development and provided both advocacy coalitions with opportunities to push for their favoured water policies.

Long-term coalition opportunity structures are a relatively recent addition to the ACF, added during the framework's most recent revision. These refer to "relatively enduring features of a polity that affect the resources and constraints of subsystem actors" (Sabatier and Weible, "Advocacy Coalition Framework" 2007, 200). In empirical applications of the ACF in European parliamentary contexts, it was found that the ACF was not sensitive enough to some of the distinctive features of these political systems, features not found in the United States. So the framework was revised to capture these distinctions, including the degree of consensus needed for major policy change, the openness of a political system, and the prevailing social cleavages, all of which may affect the resources and strategies of advocacy coalitions in a given policy subsystem. In southern Alberta, for example, it is important that the federal government has had almost no role in Alberta's domestic water governance since the early 1970s. This has created a coalition opportunity structure with only one policymaking venue – at the provincial level – negating opportunities for venue shopping and forcing coalitions into a "winner take all" contest in the provincial policy venue. Since the Aggies have a preponderance of hard power within the provincial venue, this coalition opportunity structure has been much to their advantage.

From this overview, it is clear that the ACF is a multifaceted framework that aims to capture comprehensively and meaningfully the most relevant factors shaping public policy development. However, it is important to remember that the ACF is an analytical framework, as opposed to a theory or a model. Frameworks are distinctive in the sense that they "provide a foundation for inquiry by specifying classes of variables and general relationships among them ... but they cannot in and of themselves provide explanations for, or predictions of, behaviour and outcomes" (Schlager 2007, 293). The ACF, for example, points to advocacy coalitions (and a number of other variables) as central in explaining policy development, but this alone does not constitute an explanation of policy development. The job of explanation is left to theories and models where hypotheses are developed and tested, and the supported hypotheses are formulated into working explanations (293). Theories "place values on some of the variables identified as important in a framework, posit relationships among the variables, and make predictions about likely outcomes" (296). Models are even more specific, as they isolate small parts of theories and test a limited number of variables to map the relationships between them (294–5). The relationship between frameworks, theories, and models is – ideally – a nested one, in which a framework has one or more theories within it, and, in turn, a theory has one or more models nested within it. Research and development at any one level should inform the other levels, helping to improve the overall explanatory power of an analytical approach.

After almost thirty years of research and development, the ACF exists as a fully specified framework that has spawned at least three lines of theoretical development: a theoretical line on the formation, maintenance, and dissolution of advocacy coalitions; a theoretical line on policy-oriented learning; and a theoretical line explaining major policy change. Each line has developed distinctive hypotheses, which have been empirically tested, and the results of which have informed the development of further hypotheses and the periodic revision of the framework itself. As yet, each line exists as a set of loosely connected hypotheses, rather than as fully developed and proper theories. All three theoretical lines are relevant to this study and are described in the next section, but it is the third line – the one explaining major policy change – that provides the basis for the central hypotheses investigated here to explain the shift to softer water policy in southern Alberta.

ACF Research

As described above, researchers using the ACF have been drawn to different parts of the framework and have addressed various types of research questions, resulting in three lines of ongoing theoretical development in ACF research. Some have been drawn to the concept of advocacy coalitions and have devoted considerable effort to testing and refining the concept, investigating actors' policy-relevant beliefs and their bases of collective action. Others have been drawn to the idea of policy-oriented learning, the notion that advocacy coalitions can learn, sometimes from evidence-based discussion and sometimes through negotiations with other coalitions, thereby modifying some of their secondary and policy core beliefs. Still others have been drawn to the larger question of explaining policy change in general, in which advocacy coalitions and policy-oriented learning play a role but other factors are at work as well. These three lines of ACF research are interrelated, but each also investigates distinctive hypotheses and provides different avenues of ongoing theoretical development (Jenkins-Smith et al. 2014). The three lines are evident in the ACF sub-literature on water governance, which, altogether, constitutes about twenty-five books, book chapters, and journal articles studying cases from all over the world. This section describes the three lines of ACF research, illustrating them with examples from the ACF water governance sub-literature.

The first line of research has focused on advocacy coalitions, empirically testing their existence and probing their foundations. Given the centrality of advocacy coalitions to the ACF, this line of theoretical development is fundamentally important and relevant to all ACF research. However, it is also somewhat narrow, in the sense that it focuses on only a single concept in the ACF, defining all larger research questions, such as identifying the factors underlying policy change and stability, as beyond the scope of analysis.

The research on advocacy coalitions revolves around the central ACF hypothesis that "on major controversies within a policy subsystem when policy core beliefs are in dispute, the lineup of allies and opponents tends to be rather stable over periods of a decade or so" (Jenkins-Smith et al. 2014, 195). This, in other words, is the ACF claim that advocacy coalitions exist and that they are key actors in policy development. Empirical research in a wide variety of policy fields and across a great number of polities has supported this central claim, verifying the existence of advocacy coalitions and their prevalence in policy processes (195). The importance of this empirical confirmation cannot be overstated. Without it, all other ACF research, which is based on the beliefs and actions of advocacy coalitions, would be discredited, and the framework itself would be relegated to the scrapheap of discarded policy theories.

Other research in this line has sought to specify the micro-foundations of advocacy coalitions by focusing on their two defining features, shared policy core beliefs and collective action. A number of studies have tested the claim that coalitions form around shared beliefs – a claim known as the Belief Homophily Hypothesis – and a lot of support has been found for this claim, though other factors, such as inter-personal trust, shared enemies, and perceived power, have also been found to play a role in coalition formation (Jenkins-Smith et al. 2014, 196–7). Some researchers have also tested the hierarchical belief model that is central to the ACF, distinguishing policy core beliefs from secondary beliefs, and gauging whether coalition actors show less consensus on secondary beliefs than policy core beliefs and show a willingness to give up secondary beliefs before policy core beliefs, as predicted in the model. Empirical support for these claims has been mixed, and researchers have found it difficult to distinguish clearly between secondary and policy core beliefs. A few studies have also sought to determine whether different types of coalition actors display different intensities of belief, such as whether government agencies tend to take more moderate positions within coalitions than their interest group allies, but not enough studies have been undertaken to draw any firm conclusions on these questions (195–6). The same could be said of the research on collective action, which remains an underdeveloped aspect of ACF research, despite longstanding criticism that the framework has not yet adequately accounted for the inter-actor collective action that is a defining feature of advocacy coalitions (Schlager 1995).

In the ACF water governance sub-literature, research focused specifically on advocacy coalitions features prominently. Many studies have devoted all or part of their efforts to researching the existence of advocacy coalitions in water policy subsystems. Just about all of these studies have concluded that advocacy coalitions do indeed exist, with the number of coalitions prevailing at any given time varying between one and three. Most water policy subsystems are found to have some sort of developmentalist coalition bent on exploiting water resources for economic purposes. Sometimes the developmentalists are farmers (Heinmiller 2013; Kroger 2005; Ellison 1998; King, Burkardt, and Lamb 2006; Gasteyer 2008; Menahem and Gilad 2013), while other times they are landowners (Weible and Sabatier 2009), cities (Jordan and Greenaway 1998), fishers (Weible and Sabatier 2005; Weible 2006), or some combination of economic users (Munro 1993; Zafonte and Sabatier 1998; Bukowski 2007; Smith 2009; Weible, Pattison, and Sabatier 2010; Overveld, Hermans, and Verliefde 2010; Meijerink 2005, 2008; Albright 2011; Han, Swedlow, and Unger 2014). Whatever their form, these developmentalist coalitions are usually the oldest, and sometimes the only, coalition in their respective subsystems, and they are typically opposed by environmentalist coalitions that emerged much later. Where they exist, third coalitions take a variety of forms, the most common being

First Nations coalitions (Ellison 1998; Ellison and Newmark 2010), and marketization or efficiency coalitions working to introduce market mechanisms into water policy (Jordan and Greenaway 1998; Bukowski 2007; Menahem and Gilad 2013). In short, the developmentalist versus environmentalist dynamic is quite consistent across water policy subsystems, but the presence of additional coalitions cannot be ruled out.

While there seems to be overwhelming empirical evidence of the prevalence of advocacy coalitions in water policy subsystems, the credibility of some of this evidence must be questioned. A wide variety of research methods have been used in measuring the existence of advocacy coalitions, which is, in itself, not problematic. However, the quality of these methods ranges from very rigorous and systematic (see, for example, Weible and Sabatier 2005; Weible 2006; Ainuson 2009; Weible, Pattison, and Sabatier 2010) to very loose and anecdotal (see, for example, Ellison 1998; Jordan and Greenaway 1998; Kroger 2005; Bukowski 2007; Gasteyer 2008; Smith 2009), and the overall tendency is towards the latter. In fact, many studies describe their methodology only in the most general terms, making it very difficult to gauge the veracity of their findings. The water governance sub-literature is not unusual in this regard, as Jenkins-Smith et al. (2014) have made a similar observation concerning the advocacy coalition literature, as a whole.

Overall, the takeaway lesson is that the existence (and number) of advocacy coalitions in water policy subsystems should not be taken for granted, and that rigorous methods are needed to make credible claims about coalition existence. So even though this study is more about policy change in general than advocacy coalitions specifically, it devotes considerable time and effort, particularly in chapter 4, to empirically measuring the advocacy coalitions in the Alberta water policy subsystem. Furthermore, the finding of an Aggie and, later, a Green coalition in Alberta water policy development is squarely in line with the developmentalist versus environmentalist dynamic pervading much of the ACF water governance sub-literature.

The second line of ACF research pertains to policy-oriented learning. ACF scholars have long been preoccupied with policy-oriented learning, as illustrated by the title of Sabatier and Jenkins-Smith's seminal ACF text, *Policy Change and Learning*, published in 1993. Policy-oriented learning refers to enduring changes in actor (and coalition) beliefs, which stem from experiences that have challenged and disconfirmed previous beliefs (Jenkins-Smith and Sabatier 1993). Policy-oriented learning typically occurs at the margins of coalition belief systems – at the secondary and, less frequently, policy core levels – as too much belief change too fast is likely to undermine the shared beliefs of an advocacy coalition and blow it apart.

If the ACF has any sort of normative leaning, it is probably towards the encouragement of policy-oriented learning. As indicated earlier, the framework was created to help scholars conceptualize and understand policy development in wicked problem situations. Though wicked problems are not readily solvable, the work of many ACF scholars assumes that policy-oriented learning is a positive process that can improve the management of wicked problems by drawing on new sources of (technical) knowledge and by facilitating greater stakeholder consensus. Hence, much of the research on policy-oriented learning has focused on the factors that encourage and impede policy-oriented learning, with the assumption that more learning is better.

Empirical research on policy-oriented learning has yet to confirm or disconfirm conclusively the factors that facilitate learning, but some findings in this literature are notable and relevant to this study. For example, some studies have supported the notion that policy-oriented learning is facilitated by "professional forums" in which high-level representatives from opposing coalitions are brought together, and professional norms predominate (Sotirov and Memmler 2012). A number of researchers have also found that the level of conflict between coalitions is an important factor in policy-oriented learning, though the relationship is complex: low levels of conflict impede learning for lack of motivation, and high levels of conflict impede learning as the result of hostility; but intermediate levels of conflict seem to provide the right level of motivation and congeniality between coalitions to promote idea exchange and learning (Weible 2008). Some have also proposed that the nature of the available data plays a role, such that more objective and widely accepted (often quantitative) data are more conducive to policy-oriented learning than more subjective and contested (often qualitative) data. Findings in support of this hypothesis, however, have been mixed (Sotirov and Memmler 2012). Finally, a number of studies have examined the role of policy brokers, with most finding that, where policy brokers exist, they often play a key role in laying the groundwork for cross-coalition learning (Ingold and Varone 2012).

Research on policy-oriented learning is a major part of the ACF water governance sub-literature, perhaps because of the implicit ACF assumption that policy-oriented learning offers a positive way forward in managing wicked problems. Studies by Bukowski (2007), Ainuson

(2009), Smith (2009), Meijerink (2005, 2008), Albright (2011), and Menahem and Gilad (2013) all find evidence to support the hypothesis that professionalized forums facilitate policy-oriented learning across coalitions. Meijerink (2005, 2008) also finds evidence that policy-oriented learning is facilitated by the availability of widely accepted and tractable data on water systems, and by the presence of an intermediate level of conflict between advocacy coalitions. A number of water governance studies also note the role of policy brokers, with Ainuson (2009), Smith (2009), Weible and Sabatier (2009), and Menahem and Gilad (2013) finding that brokers played an important role in bringing coalitions together and promoting learning. Weible and Sabatier (2009) further argue that a policy broker helped transform the whole tenor of coalition relations in the Lake Tahoe policy subsystem, moving it from adversarial to collaborative.

As with the research on advocacy coalitions, the quality of research on policy-oriented learning in water governance varies considerably. The most rigorous studies have relied on a combination of content analysis and process tracing, showing empirically that coalition beliefs have changed and then tracing the learning process that stimulated these changes (see, for example, Ainuson 2009; Weible and Sabatier 2009). The less-rigorous studies have relied on only one (or none) of these methods, and less confidence should be placed in these results (see, for example, Bukowski 2007; Smith 2009). The situation in the water governance sub-literature, once again, reflects the larger ACF literature on policy-oriented learning in the sense that "the most pressing concern is inconsistency in the conceptualization and measurement of the concept [i.e., learning]" (Jenkins-Smith et al. 2014, 201).

The literature on policy-oriented learning is informative to this study, even though our focus here is on policy change in general rather than policy-oriented learning specifically. Many of the factors identified in the literature as affecting policy-oriented learning feature in the development of southern Alberta water policy, as will be seen in chapters 7 and 8. The creation of professionalized forums, the availability of objective and tractable environmental data, and the presence of an influential policy broker all played notable roles in facilitating policy-oriented learning and contributing to the shift to softer water policy. These factors are part of the explanation developed in this book, but only part. Policy change can be effected through means other than policy-oriented learning, and a complete explanation of the shift to softer water policy in southern Alberta must take account of those means, as well.

This brings us to the third line of theoretical development in the ACF, which focuses on explaining policy change and stability. This is the large, overarching puzzle that prompted the development of the ACF, and it is easy to see how the first two lines of ACF research feed into this one: both advocacy coalitions and policy-oriented learning are important parts of the ACF explanation of policy change. This line of research is also the one most directly relevant to the central puzzle of this book, and the hypotheses investigated here are derived from this body of work.

ACF researchers draw basic distinctions between major policy change, minor policy change, and policy stability that is based on the framework's model of hierarchical beliefs. In the ACF, policies are assumed to be interpretations and reflections of policy-oriented beliefs. Just as policy-oriented beliefs have core and secondary aspects, so too do public policies. Core aspects of a policy have to do with its overall direction and goals, while secondary aspects of a policy have to do with the implementation of these goals. A change in the core aspects of a policy indicates a substantial change in policy direction or policy goals, and is therefore considered to be a major policy change. The shift from hard to softer water policy is a good example of such a major policy change, as the direction of policy shifted from a supply provision orientation to a demand management orientation. A change in the secondary aspects of a policy, such as minor changes in budgetary allocations, administrative rules, or legislative interpretations, deals only with implementation and does not change the direction of policy, so it is considered a minor policy change. Policy stability occurs when neither the core nor the secondary aspects of a policy are changed, leaving the status quo undisturbed (Jenkins-Smith et al. 2014, 201).

The central puzzle for ACF scholars of policy change has been explaining the underlying conditions of major policy change. Two hypotheses have been developed that aim to specify some of these conditions, each of which requires some explanation.

The first hypothesis tries to define the events that are necessary (but not sufficient) for major policy change to occur. Public policy development is a longitudinal process unfolding over time and, as such, it is substantially influenced by historical events, accumulated experiences, and learning. For instance, dramatic events, such as a natural disaster or a change in government, can alter the economic, social, or political context of policymaking, sweeping aside obstacles to reform and stimulating major policy change. Similarly, years or decades of experience and learning in a policy area will periodically reach a tipping point, changing the beliefs of key actors and opening a route to major policy change where there was none before. Students of public policy are keenly aware that these sorts of historical contingencies are quite important in explaining policy change, but because these contingencies are so difficult to predict, they are equally difficult to incorporate into theories of policy development. The ACF has addressed this challenge by recognizing four historically contingent "paths" of policy change.

Each of the ACF's four paths represents a different way in which historically contingent events can stimulate major policy change: (1) external shocks, (2) internal shocks, (3) policy-oriented learning, and (4) negotiated agreements. External shocks originate outside of a policy subsystem in the larger political, economic, social, or natural systems in which a policy subsystem is nested. When major perturbations occur in these systems, be it an election result, a drought, a flood, a recession, a species collapse, or something else, these events can dramatically affect a policy subsystem and create opportunities for policy change. Internal shocks are similar but involve a perturbation originating within a policy subsystem, such as a dramatic policy failure or the emergence of a new advocacy coalition. The third path, policy-oriented learning, occurs when a dominant advocacy coalition is confronted, often repeatedly, with evidence that falsifies some of its policy-relevant beliefs, prompting it to revise these beliefs and to make policy changes to match these revised beliefs. Finally, negotiated agreement involves compromise and negotiation between two previously warring advocacy coalitions, typically when they come to a realization that they are locked in a mutually detrimental hurting stalemate (Weible, Sabatier, and McQueen 2009, 124).

These four paths of policy change are important in the ACF because the presence of at least one of them is considered to be a necessary condition of major policy change. This notion results in the first policy change hypothesis, as expressed by Weible and Nohrstedt (2012, 133): "Significant perturbations external to the subsystem, a significant perturbation internal to the subsystem, policy-oriented learning, negotiated agreement, or some combination thereof are necessary, but not sufficient, sources of change in the policy core attributes of a governmental program."

As necessary but not sufficient conditions, at least one of the paths must be present in order for major policy change to occur. But the presence of a path does not guarantee major policy change. There are some instances where external shocks, internal shocks, policy-oriented learning, and/or negotiated agreement are present, but no major policy changes result, as we will see in chapter 6.

The first policy change hypothesis has been investigated in many ACF studies where it has, with few exceptions, found "strong support" (Jenkins-Smith et al. 2014, 203). This has also been the case in the water governance sub-literature, where many authors have used this hypothesis to explain major water policy changes all over the world. Moreover, in making these explanations, all of the policy change paths have been identified in the literature at least once, though policy-oriented learning is, by far, the most frequently identified path. There is also considerable diversity in the combinations of paths identified, with some studies identifying single paths, and other studies identifying various combinations of two or even three paths. Weible (2006) even identifies an external shock that did not result in major water policy change, reinforcing the necessary but not sufficient nature of the paths as explanatory factors.

Based on the widespread empirical support in the literature, the first policy change hypothesis seems to have great potential for helping to explain the shift to softer water policy in southern Alberta. Accordingly, it provides the basis for one of the central hypotheses of this book, namely that the policy change paths, in various combinations, were necessary factors in the shift to softer water policy. Though the paths were necessary, they were not sufficient explanatory factors, as it was left to the advocacy coalitions to exploit these paths and push for major policy change.

This leads us to the second ACF policy change hypothesis, which focuses on the role of advocacy coalitions in effecting major policy change. This hypothesis is more straightforward than the first one and simply states, "The policy core attributes of a government program in a specific jurisdiction will not be significantly revised as long as the subsystem advocacy coalition that instated the program remains in power within that jurisdiction – except when the change is imposed by a hierarchically superior jurisdiction" (Jenkins-Smith et al. 2014, 203–4).

Put differently, this hypothesis claims that major policy change occurs only when a dominant coalition is displaced, except, of course, when policy change is imposed by a higher-level government. Given the close relationship between coalitions' policy core beliefs and the core aspects of policy, this claim makes sense: a shift in the dominant coalition brings new policy core beliefs to the fore, which should produce change in a policy's core aspects.

Despite its intuitive appeal, the empirical evidence in support of the second policy-change hypothesis is not particularly strong. Some ACF studies have found evidence to support the hypothesis, but a substantial minority have not (Sotirov and Memmler 2012). Rather than finding the hypothesis to be invalid, it has generally been found to be incomplete. That is, studies have found that a shift in dominant coalition or imposition from a hierarchically superior government does generally result in major policy change, but major policy change also seems to happen in instances where these factors are not present. This is particularly evident in the water governance sub-literature, where there are many cases of major water policy change, but few cases in which developmentalist coalitions have been displaced as dominant in their policy subsystems. This means that there are a good number of cases in which major policy change has occurred, but *without* a change in the dominant coalition (and without imposition by a higher-level government), as predicted in the second policy change hypothesis (see, for example, Heinmiller 2013; Menahem and Gilad 2013; Jordan and Greenaway 1998). All of this suggests a weakness in the hypothesis and the need for some revision.

One source of weakness may be its underestimation of the role of minority coalitions in effecting major policy change. In its current form, the second policy change hypothesis does not mention minority coalitions and, as such, provides no role for them in the policy change process. Yet minority coalitions are typically the main advocates for policy change, and some minority coalitions wield substantial influence in their policy subsystems without ever becoming the dominant coalition. So it would make sense to revise the second policy change hypothesis to take greater account of the role of minority coalitions.

Another weakness in the second policy change hypothesis may stem from its apparent incompatibility with two of the policy change paths specified in the first policy change hypothesis. If major policy change can take place through policy-oriented learning or negotiated agreement, then surely dominant coalitions are capable of learning and/or compromise, and a change in dominant coalition is not necessary for major policy change. Moreover, it may very well be influential minority coalitions who are behind this learning and/or compromise, as they work to persuade dominant coalitions and policy decision-makers and pull them towards their policy core beliefs.

One way of addressing the weaknesses in the second policy change hypothesis is to focus on the inter-coalition balance of power rather than just the dominant coalition, and the concepts of hard and soft power are pertinent in this regard. Dominant coalitions are those that hold a preponderance of hard power in a policy subsystem. So a shift in the balance of hard power would signal the emergence of a new dominant coalition and likely trigger a major policy change, just as the second policy change hypothesis currently predicts. However, the balance of soft of power is also important because this is typically how minority coalitions influence policy. Thus a shift in the balance of soft power in favour of a minority coalition can also be a precursor to major policy change. When added to the existing proviso about policy imposition by higher-level governments, a new second policy change hypothesis can be stated as follows: A shift in the balance of coalition hard power, a shift in the balance of coalition soft power, imposition by a hierarchically superior jurisdiction, or some combination thereof, are necessary but not sufficient sources of change in the policy core attributes of a governmental program.

This new hypothesis is similar in style to the first policy change hypothesis as it tries to define necessary conditions of policy change. It incorporates the notion that shifts in the dominant coalition can effect major policy change (through shifts in hard power), but also recognizes that minority coalitions can effect major policy change without becoming a dominant coalition (through shifts in soft power). It also continues to recognize the accepted notion that policy changes can be imposed by higher-level governments. Moreover, the new hypothesis accommodates the real-world complexity of policy development by recognizing that these three processes – shifts in hard power, shifts in soft power, and imposition from above – may occur in various combinations. And it is important to highlight that the central concepts in the new hypothesis are readily operationalizable. As outlined above, hard and soft power can be operationalized using the six coalition resources already recognized in the ACF. This is the approach taken in chapter 4 of this volume, which is dedicated exclusively to investigating these resources and documenting shifts in hard and soft power during the period of study.

The new policy change hypothesis forms the second central hypothesis that will be used here to investigate the shift to softer water policy in southern Alberta. Since the Greens did not displace the Aggies as the dominant coalition in the water policy subsystem, and the federal government showed no inclination to try to impose water policy reform from above, these elements of the hypothesis do not factor into this study. Instead, it was a shift in the balance of soft power towards the Greens that gave them increasing influence within the policy subsystem and allowed them to pressure and persuade the Aggies and Alberta's policy decision-makers into accepting some soft water policy reforms. This shift in soft power, with little shift in hard power, also explains how the Aggies were able to defend much of the hard policy status quo while some important soft elements were added to it. The end result was a move to soft*er* water policy, an important change in policy direction but far from a complete abandonment of the longstanding policy status quo.

Summary

This chapter has examined the framework of analysis for this study, the ACF. It outlined the main tenets and assumptions of the framework, as well as the three lines of ongoing theoretical development in ACF research. The ACF conceptualizes policy development as occurring within specialized policy subsystems and involving contests between competing advocacy coalitions. Advocacy coalitions are groups of like-minded policy elites who make common cause in working to persuade government decision-makers to adopt their respective policy preferences. The number of advocacy coalitions varies in different policy subsystems and, when there is more than one, the coalitions draw on various hard and soft power resources to try to outmanoeuvre each other in influencing government decision-makers. The ACF also recognizes various kinds of "relatively stable parameters," "external subsystem events," and "long-term coalition opportunity structures" that are outside of policy subsystems, but that affect subsystem activities and have a notable impact on policy development.

There are three ongoing lines of theoretical development in ACF research: research on advocacy coalitions, research on policy-oriented learning, and research on policy change. All three lines are relevant here, but the research on policy change is the source of the two main hypotheses used in this study to investigate and explain the shift to softer water policy in southern Alberta. Put succinctly, these two hypotheses are:

1. External shocks to the water policy subsystem, internal shocks to the water policy subsystem, policy-oriented learning, negotiated agreement, and various combinations of these were necessary but

not sufficient factors in the shift to softer water policy in southern Alberta; and,

2. A shift in the balance of soft power towards the Greens was a necessary but not sufficient factor in the shift to softer water policy in southern Alberta.

Much of the remainder of this book is an empirical exploration of these hypotheses, starting with the second hypothesis in chapters 4 and 5, and then the first hypothesis in chapters 6, 7, and 8. First, however, we turn to the history of Alberta water policy in chapter 3. This chapter establishes the central premise of the study: that southern Alberta water governance has shifted from a longstanding hard policy approach to a more recent softer approach. It does so by surveying more than a century of Alberta water policy, from the 1892 Northwest Irrigation Act to the SSRB Water Management Plan of 2005.

Chapter Three

Water Policy in Southern Alberta: From "Hard" to "Soft*er*"

The water resources of Alberta are to be managed in support of the overall economic and social objectives of the Province. The Government's commitment to a program of balanced economic growth, the general welfare of Albertans, and the present and future quality of life are overriding considerations in water management. The supply of good quality water should not be a limiting factor in achieving these economic and social objectives.

Government of Alberta, *Water Resource Management Principles for Alberta*

For nearly a century, from the 1890s to the 1990s, water scarcity in southern Alberta was approached as a problem of undersupply. To correct this undersupply, governments – federal and provincial – pursued hard policies based on the construction of a water supply management infrastructure and full use of available water resources. The logic of the hard policy approach was straightforward: if water scarcity is holding back development, then more water is needed, and the way to get more water is to construct dams, diversions, and canal systems that bring more water under control and allow it to be transported to where it is needed. The focus on concrete infrastructure – often literally constructed in concrete – is what makes these policies "hard," though an implicit and important part of hard policies is a liberal approach to water allocation that allows water demands to grow, necessitating more and more water supply. By the 1960s, the water resources of southern Alberta were developed to such an extent that some hard policy advocates even began casting their eyes northward, crafting plans to divert the North Saskatchewan River into the SSRB through a massive feat of engineering, thereby solving southern Alberta's undersupply problem once and for all.

While hard policy advocates tried to engineer their way out of water scarcity, a new group of actors slowly emerged who approached the issue not as an undersupply problem but as an overuse problem. For these actors, the best way to manage scarcity was a softer, demand-side approach that restrains and reconfigures water demands to meet available supplies. Implicit in this approach is the notion that water supplies are finite and that it is neither possible nor desirable to keep finding new water supplies to meet continuously growing demands. The soft approach also involves a greater awareness of the environmental services provided by rivers and wetlands, and a greater effort to accommodate these environmental water demands. Moreover, the hard, engineered solutions pursued for decades are regarded as part of the problem rather than as part of the solution, because they interrupt, diminish, and degrade natural water flows.

Though it would be a considerable stretch to say that soft policy has completely displaced hard policy in southern Alberta, it is clear that water policy in this region has become considerably softer since the early 1990s. Since the completion of the Oldman Dam in 1992, few dams of any size have been constructed in southern Alberta, and at least one major dam proposal (the Meridian Dam) has been rejected. Even more importantly, starting with the Water Act reforms, a variety of demand-oriented water policy instruments have been introduced in the SSRB, including: water licensing moratoria to limit further growth in water demand; water licence trading to allow water to move between uses; water conservation objectives to protect and restore the SSRB's rivers; and conservation holdbacks from traded licences to allow for the recovery of modest amounts of environmental water. To be sure, the dams, canals, and water licences of the hard policy era have not been dismantled and remain central to water governance in the region. But licensing moratoria, water markets, water conservation objectives, and conservation holdbacks have been superimposed on them, creating a softer approach to water governance that is very different from that in the past.

In this chapter, we examine the evolution of water policy in southern Alberta, starting with the hard policy era from the 1890s to the 1990s and then the softer policy era succeeding it. The chapter follows a chronological order, starting with the Northwest Irrigation Act of 1894 and documenting the major water policy measures in southern Alberta throughout the twentieth century and into the twenty-first century. This brief synopsis puts the shift to softer water policy in historical context, illustrating the stark differences between hard and softer water policy, and highlighting what a significant departure the adoption of softer water policy has been in southern Alberta water governance.

The Hard Policy Era

It is a testament to the endurance of hard policy that it not only predominated in southern Alberta for about a century, but also persisted through a number of major changes in the constitutional arrangements governing water in the region. Until 1930, natural resources in southern Alberta, including water, were the jurisdiction of the federal government, and it is a piece of federal legislation, the Northwest Irrigation Act of 1894, that laid the foundations of the hard policy approach. Significant provincial involvement in water policy dates to 1915 when the Alberta legislature passed the Irrigation Districts Act, allowing farmers to form cooperative irrigation districts for constructing, maintaining, and operating irrigation infrastructure. The next sixty-five years was an era of federal-provincial cooperation as both governments worked to expand water development in the region. However, over time, the provincial role grew and the federal role shrank, until, in 1973, the federal government withdrew from its irrigation support role altogether, leaving water policy almost entirely to the provincial government. Remarkably, despite all of these jurisdictional and political changes, the one constant was adherence to hard policy.

Federal Governance

On 15 July 1870 the fledgling Canadian government – then barely three years old – took control over the area known as the Northwest Territories, a massive tract of land in the North American interior that included all of present-day Alberta, Saskatchewan, northern Manitoba, Nunavut, and the Northwest Territories. The acquisition of this vast territory was crucial to Prime Minister Sir John A. Macdonald's vision of a British dominion stretching from sea to sea across the northern half of North America. Having acquired the territory, the next task was to occupy and settle it, thereby securing it as Canadian and heading off any notions of American northward expansion. For Macdonald, this meant the construction of a transcontinental railway, located entirely within Canadian territory and linking all of the Canadian provinces and territories together. The railway would serve as an east-west transportation artery, initially for settlement and ultimately for the transport of goods in a Canadian national economy. Macdonald's vision became known as the National Policy and it dominated Canadian politics and western settlement policy even into the twentieth century.

While the transcontinental railway was the centrepiece of the National Policy, one of its underlying issues was how to develop the Prairie region once settlers had arrived. The person at the centre of the Prairie development issue was a federal public servant named William Pearce. In the 1880s, Pearce was the inspector of Dominion Lands Agencies in the Department of the Interior, a position that gave him considerable authority over the disposition of federal land, water, and mineral resources in the Northwest Territories (Mitchner 1973, 28–9). Pearce was one of the most knowledgeable people around on issues of Prairie geography and climate and he was politically well-connected within the Conservative Party. Most notably, he had the ear of Prime Minister Macdonald who asked Pearce "to report directly to him [on] any matter pertaining to western development that he deemed worthy of attention" (29). In his administrative and policy advising capacities, Pearce was given wide latitude – far more than any contemporary Canadian public servant – and he used this latitude to articulate and institutionalize a vision of water development and management that has all the hallmarks of hard water policy.

For Pearce, the southern Prairies had great agricultural potential, but the overriding constraint on development was a scarcity of water. Pearce believed that most of the southern Prairies were too dry for the kind of intensive agriculture practised in the East, and initially he saw the future of the southern Prairies as lying in ranching. Ranchers could acquire large tracts of land at relatively little cost and make use of the natural Prairie grasses as grazing fodder, supporting great herds of cattle (Mitchner 1973, 32). While never abandoning his support of ranching, Pearce also became a fervent supporter of irrigation. He believed that the successful irrigation projects of the American West could be replicated in western Canada, allowing for the introduction of intensive agriculture and thereby more intensive settlement. Irrigation would also provide fodder crops to support ranching, particularly in dry years when natural grasses were stunted and cattle stocks were threatened with starvation (35–7).

Pearce's enthusiasm for irrigation was part of his hard policy vision of water governance for the southern Prairies. He advocated for centralized government control of the region's water resources and active government involvement in the construction of a water control infrastructure that would put these water resources to productive economic use. He called for the construction of dams and water storage reservoirs in the foothills and upper reaches of the Prairie rivers, for the construction of canal and conveyance networks that would move water from the reservoirs to irrigation projects, and for support in the planning and construction of the irrigation projects themselves. He also believed that government should centrally plan and license water use in the southern Prairies, allowing for more orderly water and irrigation development than in the American West (Mitchner 1973, 41–2). Pearce was such an advocate for irrigation that he actively promoted it throughout western Canada and became a partner in one of the earliest irrigation projects in southern Alberta, all while still a public servant. When he eventually left the public service in 1904, it was for a post in the Canadian Pacific Railway, where he had a role in overseeing their irrigation projects along the Bow River (Mitchner 1973).

Pearce's vision met plenty of resistance within the Canadian government but would eventually prevail. The resistors were politicians and senior public servants in the Department of the Interior. Some simply refused to believe that the southern Prairies were water-short and relied on contemporary experts such as botanist John Macoun who claimed – falsely as it turned out – that the southern Prairies were not as dry as they first seemed and that the cultivation of the land would actually encourage greater rainfall (Jones 1987, 11–15). Others recognized the aridity of the southern Prairies but were loath to recognize it in public policy. They believed that supporting irrigation would admit the water scarcity of the region and thereby deter potential settlers, thwarting the government's western settlement ambitions (Mitchner 1973, 53). However, a prolonged drought in the late 1880s and early 1890s, together with Pearce's tireless advocacy and growing calls for irrigation by settlers themselves, combined to overcome the opposition. In the spring of 1892, Pearce was directed by the deputy minister of the interior to put together a report on his water policy recommendations for the Prairies, and that fall he was called to Ottawa to begin work on a draft irrigation bill (55–7). In this way, Pearce became the chief architect of the Northwest Irrigation Act (1894), the first major piece of water legislation in the Canadian Prairies and the cornerstone of hard water policy in the region.

The Northwest Irrigation Act introduced the prior allocation system of water entitlement distribution, a system that prevails in southern Alberta to this day. Prior allocation was a distinctly Canadian policy innovation, heavily influenced by Pearce's investigations of irrigation and water development in Australia and the western United States. The overall aim of the prior allocation system was to allow would-be irrigators ready and secure access to as much water as they could put into productive agricultural use, thereby facilitating intensive farming on the dry southern Prairies.

From the western United States, Pearce borrowed the principle of "first in time, first in right" water allocation. In the English-speaking world, the conventional approach to water allocation was based on common law riparian rights in which those owning land adjacent to a water source (i.e., riparians) have the right to take water from it. This approach worked well when water sources were abundant, but when water was scarce, as in the southern Prairies, the riparian approach restricted development to narrow bands of land along rivers, leaving plenty of potentially good land undeveloped. Governments in the water-scarce American West quickly recognized this problem and overcame it through the adoption of a "first in time, first in right" approach that allowed farmers to claim water and use it for mining or irrigation wherever they could put it to beneficial use, even on land that did not border a river. Water users could continue to claim water, as long as they put it to beneficial use, until supplies were exhausted. In dry years, when river flows were low, water was allocated on the basis of seniority of entitlements: the oldest entitlement got its water first, followed by the next oldest, and so on. This had the potential to leave junior entitlements without water in dry years, but it also provided incentive for irrigators to stake their entitlements early, thereby encouraging swift irrigation development (Tarlock 2001). Pearce recommended this seniority-based approach to prioritizing water entitlements[2] in his 1892 report, and ultimately it was accepted by Parliament.

2 In practice, Canadians have been more reluctant to utilize the seniority system during water shortages than their American counterparts and have often resorted to other means to cope with water shortages. In the summer of 2001, for instance, water licence holders in the western reaches of the Oldman sub-basin negotiated a water-sharing arrangement between them rather than calling on their licensed priorities and leaving some licence holders without water.

From Australia, Pearce borrowed the idea of government-administered water licensing. Although the Americans' "first in time, first in right" approach had its advantages, its implementation was often disorderly and conflict-ridden. Staking water entitlements in the western United States rested on irrigators' ability to prove the earliest point at which they had put a volume of water to beneficial use and was often established only through court action (Tarlock 2001). The Australian colonies, in contrast, had vested ownership of all water resources in the Crown and allocated water through government licensing, a more centralized and orderly approach that Pearce found appealing (Percy 1988, 7–8). Accordingly, under the Northwest Irrigation Act, most water rights in the Northwest Territories were vested in the Crown,[3] and most water uses required a government water licence. Water licences were distributed according to the "first in time, first in right" principle so that every licence listed a date of issue, and this date determined its prioritization relative to other licences.

The Northwest Irrigation Act, and the prior allocation system it introduced, quickly became the cornerstone of a hard water policy, because it allowed for the diversion of water to wherever it could be productively used. Access to water was no longer limited to those who owned riparian land; it now became available to anyone who could put water to a beneficial use and who could obtain a government water licence for this purpose. This opened the door to the diversion of water away from rivers, sometimes across long distances, greatly increasing the amount of water that could be used and the amount of water-based development that could be undertaken. Indeed, to a significant extent, the large and complex network of water infrastructure that criss-crosses southern Alberta owes its existence to the basic principles of the prior allocation system. Moreover, in its implementation, government control of water licence distribution was "lightly exercised," and licence "applications made in accordance with statutory requirements were rarely rejected, unless they related to a watercourse that was already fully allocated" (Percy 1988, 14). Overall, the guiding assumption was that water supply should be brought to wherever it was needed – Pearce's fundamental intention – and water policy in southern Alberta worked for decades on this basis.

---

3 The only vestigial riparian rights remaining after the Northwest Irrigation Act were rights for riparians to use water for domestic purposes.

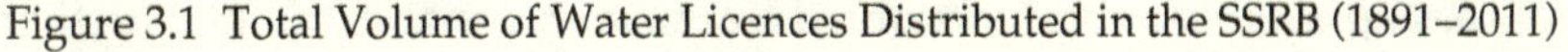

Figure 3.1 Total Volume of Water Licences Distributed in the SSRB (1891–2011)

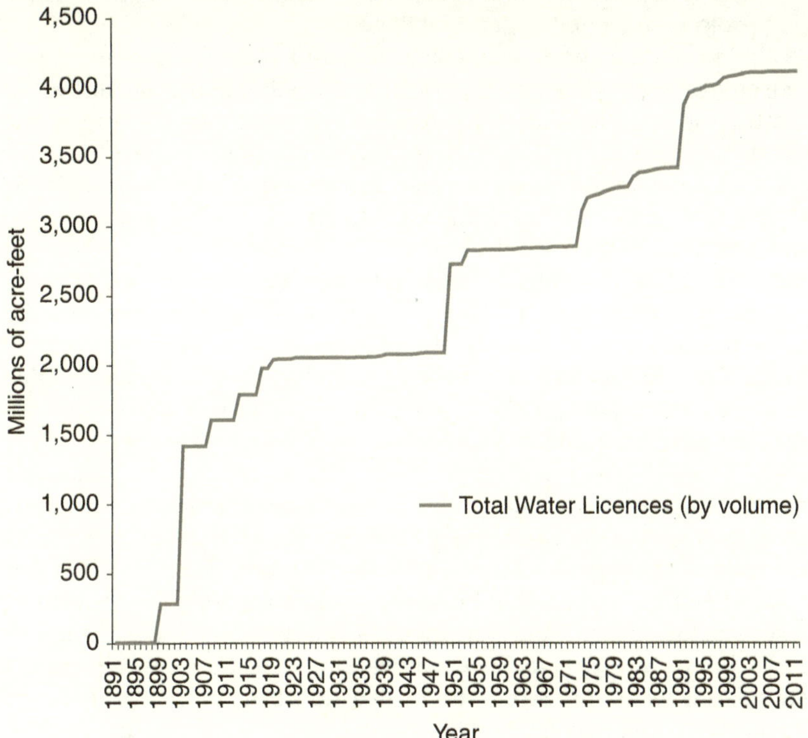

The total volume of water licences distributed in Alberta's portion of the SSRB between 1891 and 2011 is shown below in figure 3.1. It is important to note that this diagram shows the volume of licences distributed and not the volume of water actually used. Because some licences were used only partially or not at all, in most years the total volume of water actually used was lower than the total volume of licences distributed. However, examining the licensed volumes provides an interesting picture of how governments tried to steer water development, and particularly irrigation development, through water licence distribution. In the years following the passage of the Northwest Irrigation Act, for instance, there was a steep increase in the distribution of water licences as the Department of the Interior issued licences to a number of corporate and cooperative irrigation projects. However, most of these projects had very troubled starts, for reasons explained further below, and by 1920 this burst of expansion had petered out as the irrigation projects struggled to get established.

While the centralized water licensing and water management system envisioned by Pearce was institutionalized with the passage of the Northwest Irrigation Act, the extensive system of dams, storages, and canals that he also envisioned was longer in coming. Table 3.1 lists the major storage reservoirs constructed in southern Alberta, and it clearly shows that most of these dams and diversions were not undertaken until after the Second World War. The main reason for this delay was the federal government's reluctance to involve itself in such projects. Irrigation development and water control infrastructure was believed to be the preserve of those in the private sector who benefited from them. Thus, in the first half of the twentieth century, the construction of dams, weirs, and canals was left largely to private power companies, who built a number of hydroelectric dams in the upper reaches of the Bow sub-basin, and corporate irrigation projects who built small networks of dams, diversions, and canals throughout southern Alberta (Armstrong, Evenden, and Nelles 2009). These entities, and particularly the irrigation companies, struggled to find capital to build their projects and, when they managed to build water infrastructure, their rudimentary structures were often destroyed by heavy spring floods (Mitchner 1973). In this way, the Northwest Irrigation Act had the desired effect of stimulating water infrastructure construction, but it was not until active government involvement and financing of water infrastructure projects during the federal-provincial era that the second half of Pearce's vision was fully realized.

The other major issue during the period of federal governance was an international one. Two of southern Alberta's major rivers, the Milk River and the St Mary River, have their origins in Montana, so American water development on these rivers has a major impact on the downstream users in Alberta. (After flowing through Alberta for about a hundred kilometres, the Milk River flows back to Montana, making Montana both an upstream and downstream jurisdiction on the river.) In 1905, the U.S. Congress approved construction of a canal to divert water from the St Mary River to the Milk River in support of irrigation development in Montana. The Canadian government protested the canal's construction and, after having its protests ignored, threatened retaliation by approving its own diversion project that would have diverted water from the Milk River back to the St Mary

Table 3.1 Major Water Storage Reservoirs in Southern Alberta

Dam/Storage	Completed	Water source	Storage capacity (million m³)	Primary purpose
Cascade Reservoir	1912	Cascade River	114.0	Hydroelectricity
Bassano Dam and Lake Newell	1914	Bow River	173.9	Irrigation district supply
McGregor Lake	1920	Bow River and BRID Canal	330.7	Irrigation district supply
Keho Lake	1923	LNID canal	89.8	Irrigation district supply
Ghost Dam	1929	Bow River	62.1	Hydroelectricity
Glenmore Dam	1932	Elbow River	12.1	City of Calgary supply
Barrier Lake	1947	Kananaskis River	11.1	Hydroelectricity
Three Sisters Dam	1951	Spray River	72.1	Hydroelectricity
St Mary River Dam	1951	St Mary River	341.4	Irrigation district supply
Travers Reservoir	1953	Bow River	93.1	Irrigation district supply
Lower Kananaskis Lake	1955	Kananaskis River	22.9	Hydroelectricity
Upper Kananaskis Lake	1955	Kananaskis River	21.5	Hydroelectricity
Milk River Ridge Reservoir	1957	SMRID canal	97.3	Irrigation district supply
Waterton River Dam	1964	Waterton River	96.9	Irrigation district supply
Chain Lakes Reservoir	1966	Willow Creek	15.1	Irrigation district supply
Dickson Dam and Glennifer Lake	1983	Red Deer River	129.6	Water supply for various uses
Crawling Valley Dam	1983	EID canal	109.6	Irrigation district supply
Oldman River Dam	1992	Oldman River	473.1	Irrigation district supply
Pine Coulee Reservoir	1999	Willow Creek	48.1	Water supply for various uses
Clear Lake	2001	Mosquito Creek	12.3	Water supply for various uses
Twin Valley Reservoir	2004	Highwood River and Little Bow River	58.4	Water supply for various uses

Sources: Alberta Environment (*Provincial Reservoir Storage Summary* 2013), de Loë (1994, 325–6), Alberta Agriculture, Food, and Rural Development ("Irrigation Development in Alberta" 2004), Rood, Samuelson, and Bigelow (2005), Alberta Environment (n.d.).

within Canadian territory. This left the two governments in an awkward stand-off until, in 1906, the issue became absorbed in the larger negotiations towards a new and comprehensive international water governance treaty between the two countries (Simonds 1998). In fact, the St Mary–Milk diversion issue was one of two issues – the other being a conflict over hydroelectric power development in the Great Lakes Basin – that formed the backdrop for the negotiation of the landmark Boundary Waters Treaty in 1909.

The Boundary Waters Treaty directly addressed the St Mary–Milk diversion issue in Article VI, which established an international apportionment of the two rivers and gave the newly created International Joint Commission (IJC)[4] a mandate to administer it. According to Article VI, "The St Mary and Milk Rivers and their tributaries ... are to be treated as one stream for the purposes of irrigation and power, and the waters thereof shall be apportioned equally between the two countries, but in making such equal apportionment more than half may be taken from one river and less than half from the other by either country so as to afford a more beneficial use to each" (Boundary Waters Treaty 1909).

Article VI also recognized that the United States had a prior appropriation of 500 cubic feet per second (or three-quarters of the natural flow) from the Milk River and that Canada had a prior appropriation of 500 cubic feet per second (or three-quarters of the natural flow) from the St Mary River, reflecting the areas in each country where large-scale irrigation was planned or had already begun (Boundary Waters Treaty 1909). In effect, the two countries agreed to share the St Mary and Milk Rivers equitably in aggregate, but provided Alberta with a larger, prioritized share of the St Mary, and provided Montana with a larger, prioritized share of the Milk. This trade-off, it was hoped, would settle the ongoing stand-off over St Mary–Milk water diversions and allow both jurisdictions to accelerate their irrigation development.

In more than a century since Article VI's introduction, this objective has been largely achieved, though international relations on the rivers have been, at times, fractious. In the years immediately after the introduction of the Boundary Waters Treaty, conflict erupted over the interpretation of Article VI, and in 1921 the IJC was forced to issue an order

4 The IJC is a binational organization comprising three Canadian appointees and three American appointees, supported by a substantial administrative secretariat. The IJC was created as both a forum for intergovernmental coordination and exchange, and as an independent investigative, regulatory, and administrative body.

outlining an authoritative interpretation of the article. Montana, however, refused to accept this interpretation and continued to push the issue, unsuccessfully, into the early 1930s (Halliday and Faveri 2007, 81). Even as recently as 2003, Montana Governor Judy Martz undertook a campaign to have the IJC re-evaluate its 1921 order, claiming that "the Order does not equally divide the waters of the two river basins, that circumstances today are different than before 1921, and that improvements are required to the administrative procedures that implement the Order" (82). In response, the IJC held public hearings on the matter in July 2004, receiving substantial public input from a wide variety of individuals and interest groups, but issued no major changes to its order.

After devolving Crown control of water to the province in 1930 and retreating from irrigation development assistance in 1970, international relations on the St Mary and Milk Rivers became one of the few water issues in southern Alberta with any kind of federal presence. While clearly important, Article VI and the 1921 IJC order constitute the international context of domestic water governance and do not give the federal government any direct role in determining water allocation and use within southern Alberta. Since 1914, these areas have increasingly been the preserve of the provincial government.

Federal-Provincial Governance

When Alberta and Saskatchewan became provinces in 1905, they were admitted to Confederation without one crucial power held by the other provinces: ownership and control of the natural resources within their borders. Crown ownership of natural resources was retained by the federal government and would remain so until the Natural Resources Transfer Agreement of 1930.

The notion that control of the region's natural resources should remain centralized in Ottawa was largely William Pearce's. Pearce advocated centralized and planned resource development, and he argued – even before any provinces had been formed from the Northwest Territories – that provincial authority over natural resources would lead to inter-provincial resource rivalries and destructive resource mining as these young governments struggled to stimulate economic development and balance their budgets. It was much better, Pearce argued, that resource control be left with the federal government, who could develop the region's resources in a more rational and orderly fashion

(Mitchner 1973, 53). This logic was widely accepted by political elites at all levels at the time of Alberta and Saskatchewan's inception, and what little objection existed was overcome by the promise of a yearly federal subsidy to the provinces in lieu of the resource power. By 1910, however, Alberta was expressing dissatisfaction with the cash-for-resource-power arrangement, and so began twenty years of on-again, off-again negotiations between Ottawa and the Prairie provinces for the transfer of their resource powers (98–106). In the meantime, the Alberta government looked for other ways to influence water governance and development within its borders, transforming the Alberta water policy subsystem from a federal subsystem to a federal-provincial subsystem at least fifteen years prior to the finalization of the Natural Resources Transfer Agreement.

The early involvement of the Alberta government was in the area of water development, facilitating the formation and maintenance of irrigation districts through the passage of the Irrigation Districts Act in 1915. This provincial role was later formalized through a federal-provincial agreement in 1919, in which the federal government agreed to continue surveying and allocating water resources while the province agreed to continue its support of irrigation development (de Loë 1994, 109).

The Alberta government's early foray into water governance was brought about by the struggles of the early irrigation projects. The Northwest Irrigation Act was designed to give would-be irrigators ready and secure access to water, but Ottawa limited its irrigation assistance to technical support in the form of land and water surveys. Actual irrigation development was left to private enterprise (Irrigation Water Management Study Committee 2002, 9). Because constructing irrigation projects was very capital-intensive, most of the private enterprises engaged in irrigation development were large corporations, such as the Canadian Pacific Railway (CPR), or specially formed land and irrigation companies backed with money from investors in central Canada and Great Britain. Typically, their business plans involved acquiring cheap land, adding value to it with the construction of an irrigation project, then selling the land to would-be farmers at a profit and collecting yearly water delivery fees from them. Unfortunately, as the result of construction cost overruns, infrastructure failures, difficulties collecting water charges, and general inexperience with irrigation management, by 1913–14 most of the private irrigation projects in southern Alberta were either failed or failing. Given the centrality of agriculture to the developing Alberta economy, the provincial government could not allow these projects to slide into oblivion, and their response to the crisis was the Irrigation Districts Act.

The Irrigation Districts Act intended to rescue the failing private projects by helping farmers organize into local cooperatives to construct and manage irrigation projects. The districts were empowered to issue bonds to finance construction and to levy local taxes for the operation and maintenance of irrigation systems (Alberta Agriculture 2000, 4). In addition, the provincial and federal governments committed to assisting with the construction of project headworks and provided guarantees to help farmers mortgage their lands. Gradually, the cooperative district model displaced the private corporate model and, with extensive government assistance, the irrigation districts became the dominant water users in southern Alberta. The thirteen irrigation districts, listed in table 3.2, now account for about 85 per cent of irrigation in the province and hold most of the oldest and largest water licences in the SSRB (Alberta Agriculture and Rural Development 2013). This has prompted some contemporary observers to label them as the "water barons" of southern Alberta, and they are undoubtedly one of the most important policy legacies of the federal-provincial governance era.

In their early days, the irrigation districts did not fare much better than their corporate predecessors and the Alberta government soon stepped in to help. In 1921, the Alberta government offered a loan guarantee to the troubled Lethbridge Northern Irrigation District (LNID), allowing it to continue operations (Rennie 2000, 192–3). In 1926, the province again saved the LNID from bankruptcy, this time taking on its debt and administering the district through a provincially appointed trustee, an arrangement that persisted until 1968. As the CPR gradually lost interest in its money-losing irrigation projects in the late 1930s and 1940s, these assets were sold to the Alberta government or to the project farmers, most of which eventually came to the Alberta government for financial assistance (Glenn 1999, 21–3). By the early 1950s, provincial ownership and administration of irrigation districts, many operating as Crown corporations, had become the norm (Prairie Farm Rehabilitation Administration 1982, 15–34; Raby 1965). Most of the provincial assistance was dedicated to keeping the irrigation districts viable and enabling them to use all of the water for which they were already licensed. That is why, despite substantial provincial assistance, the distribution of new water licences experienced a prolonged plateau from the early 1920s to the late 1940s, as shown in figure 3.1.

Table 3.2 Alberta's Thirteen Irrigation Districts

Irrigation districts	Date formed	Assessed area (hectares)
Taber Irrigation District (TID)	1917	33,178
Lethbridge Northern Irrigation District (LNID)	1919	62,680
United Irrigation District (UID)	1921	13,902
Mountain View Irrigation District (MVID)	1923	1,506
Raymond Irrigation District (RID)	1925	18,515
Magrath Irrigation District (MID)	1926	7,406
Eastern Irrigation District (EID)	1935	113,760
Leavitt Irrigation District (LID)	1936	1,930
Western Irrigation District (WID)	1944	35,665
Aetna Irrigation District (AID)	1945	1,461
Ross Creek Irrigation District (RCID)	1949	490
St Mary River Irrigation District (SMRID)	1968	148,585
Bow River Irrigation District (BRID)	1968	85,450

Source: Irrigation Water Management Study Committee (2002).

Two decades after the province became deeply involved in irrigation assistance, so too did the federal government, primarily through the work of the Prairie Farm Rehabilitation Administration (PFRA). The PFRA was created in 1935 in the midst of the Great Depression and the Prairie Dustbowl with a mandate to preserve and improve the viability of Prairie agriculture. The PFRA identified irrigation as a priority area and directed massive sums of money towards irrigation development (Gray 1996). The Alberta government, which had been virtually bankrupted by the Depression, accepted the assistance and worked cooperatively with the PFRA in its development efforts. Over the course of forty years, the PFRA constructed water control infrastructure to supply the irrigation districts, and also played an important role in rehabilitating and expanding the St Mary Irrigation District and the Bow River Irrigation District, two of the largest irrigation districts in the province (Irrigation Water Management Study Committee 2002, 16). Between 1935 and 1978, the PFRA spent around $600 million on water control infrastructure on the Canadian Prairies, much of it directed to southern Alberta (Prairie Farm Rehabilitation Administration 1982, 13). The effects of the PFRA's efforts can be seen in figure 3.1 as a spike in new water licences occurred in the early 1950s and again in the early 1970s as PFRA projects came online and made new water supplies available to irrigators.

By the 1920s, it became increasingly clear that federal control of Alberta's and Saskatchewan's natural resources was a constitutional anachronism in need of correction, and federal-provincial negotiations on the subject intensified. The Prairie governments argued that they were being denied a crucial instrument of economic development and demanded equal treatment with the other provinces, which had always had control of the natural resources within their borders. The result was the negotiation of the Natural Resources Transfer Agreements in 1930. While it was assumed by all involved that this transfer included water resources, some doubt about this arose in the years immediately after the agreements, and in 1938 the agreements were amended to clarify that control of water resources had indeed been part of the transfer (Percy 1988, 11). The Natural Resources Transfer Agreements marked a significant federal retreat from water governance in southern Alberta. From this point forward, Ottawa's influence in the southern Alberta water policy subsystem was exercised less through regulatory means and more through its spending power, largely in the activities of the PFRA.

Even with a new regulator in charge, the prior allocation system continued virtually unchanged. Alberta's 1931 Water Resources Act was a near carbon copy of the 1894 Northwest Irrigation Act, continuing the practices of Crown water ownership, water access through government licensing, and licence prioritization on the basis of seniority. All of the licences distributed under the Northwest Irrigation Act were preserved and their original priority dates recognized under the new provincial regime (Percy 1996–7). In effect, the hard policy introduced by the federal government was continued by the Alberta government and would remain largely unchanged for a further sixty-five years.

The transfer of Crown control over water also introduced the possibility of inter-provincial water allocation conflicts, as William Pearce had predicted decades earlier. Alberta was the upstream jurisdiction on most Prairie rivers, and the downstream provinces of Saskatchewan and Manitoba sought some assurances that Alberta would not capture and use all of this water before it had a chance to reach them.

The discussions over inter-provincial water management on the Prairies were long and difficult, spanning almost forty years, but the apportionment arrangements finalized in 1969 have proven stable and generally acceptable to all involved. In 1930, in conjunction with the Natural Resources Transfer Agreements, Alberta, Saskatchewan, and Manitoba formed the Western Water Board, a coordinative body for addressing shared water issues. This was replaced in 1948 by the Prairie Provinces Water Board (PPWB), which included not only provincial but federal representatives. The PPWB was created to recommend annual water apportionments between the three provinces, but this approach broke down in the 1960s as the provinces began asking for larger and larger apportionments. Negotiations were then undertaken towards a permanent apportionment of the Prairie rivers resulting in the 1969 Master Agreement on Apportionment (Prairie Provinces Water Board 2004).

Like Article VI of the BWT, the Master Agreement on Apportionment is an important set of rules setting the context for water governance in southern Alberta. The agreement "contains a simple formula based on the principle of equal sharing of available water in the Prairies" (Prairie Provinces Water Board 2004). For Alberta, this means that it is entitled to use half of the natural water flow in its rivers, as measured at the inter-provincial border, allowing the other half to flow into Saskatchewan. Crucially, the two major inter-provincial rivers in the southern Prairies, the Red Deer and the South Saskatchewan, are treated by the agreement as one river (Master Agreement on Apportionment 1969). This has meant that Alberta can take more than half of the flow from the South Saskatchewan – which it periodically does – as long as it compensates with extra flow from the Red Deer (Figliuzzi 2002, 11). This has allowed water development to continue unconstrained in the southernmost parts of the SSRB and has served as a means of reconciling Alberta's apportionment commitments with its hard approach to water governance. The Master Agreement on Apportionment is overseen by the PPWB, which includes a federal representative and relies on water monitoring stations operated by Environment Canada. However, neither the PPWB nor Ottawa has enforcement powers to force the provinces to comply with their apportionment commitments (Prairie Provinces Water Board 2004).

By the early 1970s, irrigation development in southern Alberta was no longer a priority for the federal government. The Prairie Dust Bowl of the 1930s was a distant memory and the Trudeau Liberals' attention was directed elsewhere. In 1973, the federal government formally withdrew from its irrigation support activities, negotiating an agreement with the Alberta government to transfer its remaining interests in the irrigation infrastructure of the Bow River and St Mary River irrigation districts (Irrigation Water Management Study Committee 2002, 17). The PFRA lived on, but federal presence in the southern Alberta water policy subsystem was greatly diminished.

Since 1973, the federal government has been a peripheral figure in southern Alberta's water allocation and water development issues.

It remained involved in the international and inter-provincial water apportionment institutions for the Prairie rivers, and it was still relevant in water issues pertaining to navigation, shipping, and fisheries, all matters clearly in federal jurisdiction under section 91 of the Constitution Act, 1867. Ottawa also maintained a role in water monitoring and research and, in the early 1990s, reluctantly applied its environmental assessment process to the proposed Oldman River Dam. For most of this period, though, the federal government preferred to take a minimalist approach to its water-related powers in southern Alberta, leaving the core water allocation and development issues almost entirely to the province. This has been especially true since the federal-provincial resource conflicts of the 1970s and 1980s, when control of natural resources became a real sore point in Alberta-Ottawa relations. Thus, in all practical senses, the southern Alberta water policy subsystem has been a provincial subsystem since the early 1970s.

Crucially, the federal government's retreat from Alberta water policy rules out one potential explanatory factor specified in the second ACF policy change hypothesis: policy change due to imposition by a higher-order government. With Ottawa out of the picture, it is clear that the shift to softer water policy in Alberta was *not* a federal project imposed on Alberta from the outside. Instead, the policy shift was the result of factors internal to the provincial water policy subsystem. More specifically, the second policy change hypothesis suggests that these factors involved a shift in the coalition balance of power, whether it was the balance of hard power, the balance of soft power, or both. We will pick up this line of investigation further in chapter 5.

Provincial Governance

The 1970s was a decade of unprecedented irrigation growth in southern Alberta. From 1970 to 1979, the area of assessed irrigable land in the SSRB grew by 50 per cent, from 279,877 hectares to 419,730 hectares. This growth can also be seen in figure 3.1 through a spike in new water licences starting in the mid-1970s. One reason for this growth was technological. Pump and sprinkler irrigation systems had become commercially viable in the late 1960s, meaning that farmers no longer had to rely only on gravity-fed irrigation systems. Any farmer who could afford a pump and who could get access to a water source was now a potential irrigator. In addition, hilly land that could not easily be irrigated through gravity was now accessible to irrigation with pumps and sprinklers. Even more important than pump and sprinkler technology, though, was the provincial government's adherence to a hard water policy that aimed to use this technology, and the water supply in southern Alberta, to its fullest.

When the federal government vacated its irrigation support role, the province moved in immediately to maintain and even expand public assistance for irrigation development. Starting in 1973, the province adopted a policy of acquiring and operating the headworks of the irrigation districts at no cost to them, deeming the operation of this infrastructure to be in the provincial interest. The Lougheed government also introduced a vastly expanded program to refurbish and expand the aging water infrastructure within the irrigation districts, with 86 per cent of the costs paid by the province and 14 per cent paid by the districts (de Loë 1994, 208–9; Irrigation Water Management Study Committee 2002, 17).

In order to provide the irrigation districts and others with a reliable water supply, the Alberta government also considered a number of large-scale dam and diversion options to give them more control over water flows in the SSRB.

One of these options was the Prairie Rivers Improvement, Management, and Evaluation (PRIME) scheme, an inter-basin diversion project designed to transport water from the water-rich North Saskatchewan River Basin to the water-poor SSRB, thereby augmenting supplies in the south and allowing irrigation expansion to continue unabated. PRIME was the pet project of the Water Resources Division of the Alberta Department of Agriculture and was modelled on large-scale inter-basin diversion schemes already being constructed in places like California and Australia. Between 1965 and 1970, the Water Resources Division investigated potential reservoir sites for PRIME and conducted various other feasibility studies (de Loë 1994, 198). The scheme was ultimately rejected by Premier Lougheed in 1971, partly on environmental grounds, but largely on the basis of public fear that PRIME would be the first step towards large-scale water exports to the western United States (200, 208). PRIME continues to have supporters in southern Alberta, and discussion of the scheme resurfaces during every drought, but its construction now seems very unlikely, given the major political changes documented in this book (Struzik and Brooymans 2002).

While PRIME was shelved, a somewhat less-radical plan for a large storage dam on the Oldman River became the favoured water supply management option and was adopted as government policy in 1975.

Plans for a dam on the Oldman River Dam dated back over fifty years, but by the mid-1970s, a combination of growing water supply shortages in the southern part of the SSRB and politicking by the Lougheed government seeking to attract southern voters made the dam a reality. The Oldman Dam was a serious test of the Alberta government's commitment to hard water policy. It was the largest and most expensive water project ever undertaken by the province, it met unprecedented criticism and resistance from First Nations and environmentalists, and, by the early 1990s, the dam was unpopular with the Alberta public. Even after a federal environmental assessment panel recommended that the dam be decommissioned in 1992, the Alberta government stood by the project and brought it to fruition (Glenn 1999). This was the last time the Alberta government showed such commitment to a hard policy initiative and, in this sense, the licensing of the Oldman Dam in August 1992 marked the end of the hard policy era in southern Alberta.

The initial softening of Alberta's water policy can be traced to the late 1970s as government officials began to view water scarcity in southern Alberta as a problem of overuse – at least in part – and took the first measures to restrain the region's ever-growing water demand. In 1978, Alberta Environment released its Water Management Principles for Alberta, a policy document that consolidated the government's water policy and outlined its vision of water management. It reaffirmed the government's longstanding commitment to the hard policy goals of managing and augmenting the water supply in support of irrigation and other economic uses. Yet it also included commitments to environmental protection, conservation, and reduction of consumption, distinctly softer policy goals somewhat at odds with longstanding practice (de Loë 1994, 210–11). The 1978 policy document was followed up in the 1980s with the South Saskatchewan River Basin Planning Program, a decade-long planning exercise that resulted in the introduction of expansion limits for the irrigation districts and minimum instream flow objectives for the Waterton, Belly, and St Mary Rivers (208–9). Though the expansion limits were clearly styled as soft policy measures, environmentalists criticized them as being much too high and the minimum instream flow objectives as being much too low. These criticisms, combined with the government's stubborn support of the Oldman Dam, illustrate the extent to which the hard policy approach prevailed, even into the early 1990s. We will explore the politics and process behind the Oldman Dam, the Water Management Principles for Alberta, and the South Saskatchewan River Basin Planning Program in further detail in chapter 6.

The Soft*er* Policy Era

Until the 1990s, the policy measures adopted to manage water scarcity in southern Alberta focused overwhelmingly on managing and augmenting the available water supply. Instead of constraining water demand, governments actively encouraged water demand growth, especially in the irrigation districts, taking growing water demand as a sign of economic progress. The most common demand-oriented (or soft) policy instruments that can be used to send conservation signals to water users were either under-used or not used at all. Water licences, for example, were not fungible, so there was little incentive for licence holders to invest in water conservation measures, such as replacing leaky infrastructure or upgrading to more water-efficient irrigation technologies, because licences were cheap but conservation measures were not. Similarly, most water users did not pay anywhere near the full-cost of the water they used because the cost of water deliveries was heavily subsidized by government, particularly in the irrigation districts. There were also few limits on the distribution of new water licences, and there was a general expectation that the government would make water available when and where it was needed.

While hard policy predominated, there were some early soft policy instruments, mostly designed to deal with instances of acute local scarcity. The earliest of these instruments was adopted through an amendment to the Northwest Irrigation Act in 1920. The amendment created a list of preferential water uses, listing domestic use as the highest priority use followed by municipal, industrial, irrigation, and other uses, respectively. When a river or stream became fully allocated and no new water was available, anyone requiring water for a higher priority use could apply to acquire or expropriate water from a lower priority use. This created something of a "safety valve" for coping with scarcity when hard solutions were not immediately available, but it was used quite sparingly before being abandoned altogether in the Water Act reforms (Percy 1996–7, 225). Another early soft policy instrument was the attachment of "staging conditions" to water licences, restricting when diversions could take place on the basis of river flow conditions. Early licences tended to have relatively few such conditions, but later licences became more restrictive as regulators became increasingly cognizant of the importance of preserving instream flows. Both the hierarchy of uses and staging conditions are notable for their efforts to shape and constrain water demand, but they were also minor appendages in the overall hard policy approach.

The shift to softer water policy really began with the passage of the Water Act in 1996. The Water Act introduced a number of new soft policy instruments – including water licensing moratoria, water licence trading, water conservation objectives, and water conservation holdbacks – which have come to play a central role in how water is governed in southern Alberta. At the same time, the number and size of dams constructed in the region fell dramatically, indicating a shift in emphasis by the Alberta government from hard policy solutions to soft ones. Nevertheless, the water supply infrastructure and the prior allocation system from the hard policy era have not been abandoned. Instead, the new soft policy measures have been overlaid on the existing hard policy measures similar to what Streeck and Thelen (2005) call "institutional layering" (22–4). Though the change has been sedimentary rather than sweeping, the cumulative change has been quite significant, particularly in a policy subsystem where the hard policy approach had prevailed for about a century.

The Water Act makes a crucial distinction between existing licences and new licences, and the two classes are treated very differently. Existing licences are those that existed at the time the legislation was passed, while new licences are those that have been issued since the legislation took effect. The Act states that existing licensees "can continue to divert water in accordance with their original priority, the terms and conditions of their original licence and the new Act. However, if there is a conflict between a term of a deemed licence and the new Act, the term of the licence prevails over the Act" (Percy 1996–7, 229). This, in effect, protects existing licences from some provisions of the Water Act that might threaten them and their water allocations. In contrast, new licences distributed under the Water Act are subject to fixed terms and can be denied on the grounds that they contravene an approved water management plan (discussed below) or would negatively affect a riverine environment (237–8). The Water Act also provides its director – a senior administrator in Alberta Environment – the authority to impose moratoria on new water licences in heavily allocated streams and empowers the environment minister to reserve unallocated water for any purpose, including the preservation of instream flows (ibid.). Altogether, these soft policy instruments provide considerable capacity to halt, or at least slow, the growth in water demand that was exacerbating scarcity in many parts of southern Alberta.

Perhaps the greatest change brought about by the Water Act was the introduction of water licence trading. The Act permitted both temporary and permanent transfers of water licences, subject to some conditions. It was envisioned that water trading would be introduced on a basin-by-basin basis and would be limited to basins where water was already scarce, such as the SSRB. Accordingly, the Act called for the creation of basin water management plans, which, among other things, were to include specific rules for water trading, developed through public and stakeholder consultations and approved by Cabinet. It was through the basin water management plans that regular, routinized water trading would be introduced. In the absence of such plans, water trading was permitted, but each trade was subject to Cabinet approval (Percy 1996–7, 236). In 2002, the first part of a basin water management plan was approved for the SSRB, allowing water trading to begin there in earnest.

In conjunction with water licence trading, the Water Act also introduced conservation holdbacks as a policy instrument for recovering water for environmental purposes. A conservation holdback is a bit like a water tax on a water trade: it is a small volume of water that is held back from a water trade and reallocated for environmental use. Specifically, the Water Act empowers its director to withhold up to ten per cent of the volume of any water traded in order "to protect the aquatic environment or to implement a water conservation objective, if the holdback is authorized in an approved water management plan" (Percy 1996–7, 239). This is the only provision in Alberta's water law that allows for any amount of water to be taken from existing licences and returned to the environment, and this instrument is central to the achievement of the water conservation objectives outlined in the basin water management plans.

Water conservation objectives are essentially instream flow targets, formalized in basin water management plans, and intended to maintain or restore healthy riverine environments. All new water licences are distributed on the condition that they do not impair the achievement of a water conservation objective. In this way, the new water conservation objectives have the potential to limit water-takings on streams that are not already fully allocated, such as the Red Deer River. Water conservation objectives have no impact on existing licences, except when existing licences are traded. During water trades, they serve as an important benchmark in the director's decision to take conservation holdbacks: holdbacks are much more likely in sub-basins with unmet water conservation objectives.

The Water Act created a variety of soft water policy instruments, but the implementation of many of these instruments depended on the formulation of basin water management plans. Of all the basins in Alberta, the priority for the Alberta government was the formulation of a plan for the SSRB, where water scarcity was being felt most acutely. From 2000 to 2005, over two phases involving extensive stakeholder consultations and negotiations, a basin water management plan was negotiated for the SSRB, and this plan is now the centrepiece of the new, softer approach to water governance in southern Alberta.

In totality, the SSRB Water Management Plan effectively introduced a cap-and-trade water governance regime in the basin. The plan introduced moratoria on new water licences in three of the SSRB's four sub-basins: the Oldman, the Bow, and the South Saskatchewan. All unallocated water in these sub-basins was reserved to the Crown and made ineligible for licensing, except for a few water projects that were grandfathered in (Province of Alberta 2007). A licensing moratorium was not introduced in the remaining sub-basin, the Red Deer, but the water conservation objectives adopted for this sub-basin effectively imposed a ceiling on the distribution of new water licences: new licences can be distributed only up to the point that the water conservation objectives can be met, and not beyond that. So, in effect, licences were capped at current licensing levels in the Oldman, Bow, and South Saskatchewan, and capped at a yet-to-be-reached level in the Red Deer. Water licence trading was permitted throughout the entire SSRB, and the director's power to impose conservation holdbacks on all water trades was affirmed. In this way, a cap-and-trade regime was superimposed on the prior allocation regime and water management infrastructure inherited from the hard policy era, shifting water governance in the SSRB in a substantially softer direction.

At the same time that it was pursuing the SSRB Water Management Plan, the Alberta government also introduced a more general water policy initiative entitled Water for Life that had a decidedly softer bent. Developed between 2001 and 2003, and involving extensive public and stakeholder consultations, Water for Life had three basic policy aims: to ensure a safe and secure drinking water supply; to maintain and restore healthy aquatic ecosystems; and to achieve a reliable water supply for a sustainable economy. What distinguishes Water for Life as a soft policy is its overwhelming reliance on research, monitoring, planning, public education, and partnerships with stakeholder groups in order to achieve its objectives (Government of Alberta 2003). The overriding emphasis is on understanding and improving Albertans' conservation of existing water supplies, rather than finding new water supplies to allow water demand to grow further. Water for Life did not alter the prevailing water policy in southern Alberta, but it was an important complement to it. The initiative proved so popular with the public and politicians that it was renewed for another five years in 2008.

The shift to softer policy since the mid-1990s has been evident not only in the introduction of new soft policy instruments, but also in the relative decline of old hard policy instruments. Since the completion of the Oldman Dam in 1992, only two new government dams have been constructed in southern Alberta: Pine Coulee Dam and Twin Valley Dam (Rood, Samuelson, and Bigelow 2005). Both dams are relatively small, together constituting only 6.2 per cent of provincially owned water storage capacity in the SSRB. Moreover, the Pine Coulee Dam had been in the works and had received partial approval prior to 1992, making it mostly a product of the hard policy era (ibid.). In 2002, the governments of Alberta and Saskatchewan investigated the construction of the proposed Meridian Dam on the mainstem of the South Saskatchewan River near the provincial border, but ultimately rejected the proposal (Canadian Broadcasting Corporation 2002). Since 2004, and the completion of the Twin Valley Dam, no new government dams have been constructed in southern Alberta, and new dam starts seem to have fallen off the water policy agenda. This is a remarkable shift in policy, considering that, only two decades previously, the Alberta government persevered through protests, negative public opinion, and an unfavourable federal environmental assessment in order to see the Oldman Dam through to completion.

The overall effect of the shift to softer policy can be seen in figure 3.1 in the plateau in new water licences since the early 2000s. This plateau can be accounted for by the licensing moratoria imposed in the Oldman, Bow, and South Saskatchewan sub-basins. New water users in these areas are now being accommodated through the water market rather than through the distribution of new licences. Some new licences have been distributed in connection with the Pine Coulee and the Twin Valley Dam projects, and to a small number of First Nations, explaining why the plateau in water licences has not been entirely flat over the last fifteen years (confidential interview 12 February 2012).

Summary

This chapter has shown how southern Alberta water policy has undergone a significant shift from hard to softer policy since the early 1990s.

Evidence of a softer approach to water governance can be found as early as 1978, but it was not until the mid-1990s, in the wake of the Oldman Dam conflict, that the softer approach really started to take hold. Starting with the passage of the Water Act in 1996 and culminating in the formation of the SSRB Water Management Plan of 2005, the Alberta government adopted a variety of softer policy instruments, while the number of new dam starts dropped dramatically and dam construction seemed to lose its appeal as a water policy instrument. It is important to note, however, that the Alberta government did not abandon the prior allocation system or the extensive water management infrastructure system that it (and the federal government) had built up over the previous century. Instead, the new softer policy measures were overlaid on these hard measures in a process of institutional layering.

The remainder of this book is dedicated to exploring and explaining the politics behind this shift from hard to softer water policy. In this effort, the next chapter explores the prevailing policy-relevant beliefs in the southern Alberta water policy subsystem from the mid-1970s to the mid-2000s. It also investigates the actors holding these beliefs and how they worked together – in advocacy coalitions – to see their beliefs reflected in policy. Establishing these actors, and their beliefs, uncovers the first piece in the puzzle of the shift to softer water policy.

Chapter Four

Advocacy Coalitions in the Alberta Water Policy Subsystem

Any arguments against the construction of dams for irrigation come from people who have never had empty stomachs.

George Dudley, Magrath Irrigation District, 1978

For decades, water managers and engineers in Alberta dreamed of water mega-projects that would harness the province's large northern rivers and divert them south to support irrigation. The earliest mega-projects were investigated in the early 1920s by the government of Canada. Two proposals were developed to divert some of the headwaters of the North Saskatchewan Basin to the south and east through a series of rivers and canals, ultimately irrigating between 630,000 and 720,000 hectares of land (de Loë 1994, 124–5). The proposals were shelved in 1923 because they were enormously costly, but they were revived by the PFRA in 1935, and again by the province in 1958. In 1963, the Water Resources Branch in Alberta's Department of Agriculture began developing an even more ambitious water mega-project, aimed at diverting more northern rivers, bringing more water south, and irrigating even more land. This was the PRIME project described in previous chapters, and it persisted as a serious water policy option into the 1970s and beyond (163).

So, for about fifty years, inter-basin diversions were an important part of Alberta's water policy discussions, and many policy elites believed that these mega-projects were the logical next step in Alberta's water management efforts, once the water resources of the SSRB had been fully tapped. Irrigators, hydrological engineers, and even many politicians greeted these inter-basin diversion proposals with considerable enthusiasm every time they came up for discussion. By the 1960s, construction of comparable inter-basin diversion projects was already underway in Australia, California, and elsewhere, so a mega-project in Alberta seemed quite viable. Moreover, a vast network of canals already moved water all around southern Alberta, so PRIME was seen as simply an expansion of this network. The only thing holding it back was its massive cost, but, as the water resources of the SSRB became increasingly depleted, proponents argued that there were few other options for dealing with the South's water scarcity and that the large costs were an investment in the province's future.

Fast forward to the mid-1990s, and things had changed dramatically. Rather than celebrating the construction of PRIME, provisions in the Water Act banned it and all future construction of inter-basin diversions. The costs of such diversions were now a political non-starter, and most water policy elites objected to inter-basin diversions on environmental grounds alone. The depletion of northern rivers, the dramatic alteration of natural aquatic environments, and the transformation of natural Prairie landscapes simply could not be justified in the name of further irrigation expansion, except by the most ardent irrigation boosters. Rather than investments in the future, inter-basin diversions came to be seen as dystopian monstrosities that had to be avoided. What accounts for this dramatic about-face?

The short answer is changing policy-relevant beliefs. Between the mid-1960s and the mid-1990s, the substance of the inter-basin diversion proposals did not really change, but the ideological prisms through which policy elites viewed these projects did change. Social changes such as the rise of environmentalism meant that inter-basin diversion projects that had once been regarded as practical infrastructure investments were now viewed as economically draining, socially disruptive, and environmentally destructive. Accordingly, projects that had once been seriously discussed in Cabinet were, thirty years later, effectively banned as a matter of law.

These sorts of changes in policy-relevant beliefs are the subject of this chapter and are the starting point for our investigation of Alberta's shift to softer water policy. Policy-relevant beliefs are a foundational variable of the ACF, playing a crucial role in the formation of advocacy coalitions and other processes such as policy-oriented learning. The chapter begins with an explanation of the place of policy-relevant beliefs in the ACF, before turning to an empirical investigation of policy-relevant beliefs and advocacy coalitions in the southern Alberta water policy subsystem between 1978 and 2005.

The ACF and Policy-Relevant Beliefs

As outlined in chapter 2, the ACF views policy development as a contest between contending advocacy coalitions in a policy subsystem. Since each advocacy coalition represents a distinct and relatively coherent set of policy-relevant beliefs, policy development is also a contest between contending sets of policy-relevant beliefs. Changes in policy-relevant beliefs are important as they contribute to the rise and decline of advocacy coalitions and to processes of policy-oriented learning.

The ACF's fundamental claim that "beliefs matter" may seem self-evident, but it actually sets the ACF apart from other mainstream theories of the policy process. For instance, it stands in sharp contrast to rational choice theories of policy development, which assume that policy actors are fundamentally self-interested and pursue policies that maximize their self-interest. The chief difference between the ACF and rational choice is the way in which each approach ascribes policy actors' motivations: in rational choice analyses, motivation (conceptualized as self-interest) is treated as something that can be exogenously calculated and ascribed to actors, while in ACF analyses, motivation (conceptualized as policy-relevant beliefs) is treated as something that must be endogenously discovered and measured before it can be ascribed to actors. This is not to suggest that self-interest and beliefs are entirely separate and mutually exclusive. There may be a good deal of self-interest informing actors' beliefs, but this is not particularly problematic to the ACF. Whether policy-relevant beliefs are based on material self-interest, some set of non-materialist values, or a bit of both, beliefs are always treated as a key variable to be measured, and this is the task taken up in this chapter.

Of the three levels of policy-relevant beliefs recognized in the ACF – those being deep core, policy core, and secondary beliefs – the focus of this chapter is on policy core beliefs. Policy core beliefs are important in the ACF because they are prevalent in the formation, maintenance, and decline of advocacy coalitions. Common policy core beliefs are crucial to an advocacy coalition because they serve as both a program to be implemented and as a rallying point for collective action. Indeed, recent research investigating the origins of advocacy coalitions found that "perceived belief similarity constitutes the major driving force behind coordination and coalition formation" (Matti and Sandstrom 2013, 252). In other words, like-mindedness does not guarantee collective action, but it is almost certainly a prerequisite for it.

Policy core beliefs are also important in the ACF because changing policy core beliefs are a preliminary indicator of policy-oriented learning, one the four "paths" of policy change currently recognized in the ACF. As shown in figure 2.1 the ACF explicitly incorporates feedback loops from policy outputs and policy impacts to advocacy coalitions and their policy-relevant beliefs, and it is through these feedback loops that actors and coalitions have the opportunity to learn and adapt, over time. They may gradually learn, for instance, that some of their preferred policy options are not technically feasible or politically viable and, after some introspection, modify their policy core beliefs to adapt to these new realities. Policy core beliefs are not readily or easily changed, and it may take a number of lessons and a considerable amount of time before coalitions adapt their policy core beliefs through learning. Moreover, most learning of this sort is marginal in scope, as too much belief change too fast can threaten a coalition's very existence. Changing policy core beliefs are usually a good indication of policy-oriented learning, but one must also be able to point to a learning event to claim with any certainty that policy-oriented learning has indeed occurred.

Given the importance of policy core beliefs in the ACF, it is not surprising that ACF scholars typically go to great lengths to measure the prevailing policy core beliefs in the policy subsystems they study. ACF scholars have gathered data on policy core beliefs using one (or more) of three data collection techniques: interviews with policy elites (Ingold 2011), surveys of policy elites (Weible and Sabatier 2009), and/or textual analysis of policy elites' documents (Nohrstedt 2008; Pierce 2011). Since the period of study in this book stretches back into the early 1970s, interviews and surveys were simply not viable. Fortunately, thousands of pages of testimony and submissions were available from policy consultation processes in 1978, 1984, 1995, and 2002/5 (the last was two phases of a single policy process). This made a textual analysis possible, and a qualitative content analysis was undertaken to gain a complete picture of the relevant actors and policy core beliefs throughout the period of study. A full description of the content analysis methodology is provided in the appendix, but a couple of key features are worth mentioning here.

The sampling procedure, used to identify a sample of useful material from the thousands of pages of submissions available, is important because it identified those actors who were most relevant in water policy development. The ACF assumes that the most relevant policy actors are those who participate regularly in policy development processes. These actors have the most knowledge of a particular policy area, have the most highly developed sets of policy-relevant beliefs, and are the most likely to organize themselves into advocacy coalitions. Thus, in determining the actors most relevant in policy development, "the ACF emphasizes frequent participants in policymaking, who are differentiated from one-shot participants that only participate on single occasions" (Nohrstedt 2011, 468). The sample of documents included in the content analysis was constructed on this basis. All submissions from actors who participated in at least two of the four policy consultation processes were included, while all submissions from one-shot participants were excluded. Accordingly, the actors included in the sample, and listed in tables 4.1, 4.2, and 4.3, were purposively selected as the most relevant participants in Alberta water policy development, and this alone provides important information about the policy subsystem in 1978, 1995, and 2002/5, respectively.

The central concept (or coding unit) of the content analysis was policy recommendations, which was used a proxy for policy core beliefs. This meant that the actor submissions were analysed, first, to identify and isolate all of the policy recommendations in the documents, and, later, to code these recommendations systematically. Similar to the approach taken by Leifeld (2013, 173–4), policy recommendations were conceptualized as having two components: a policy concept and a policy disposition. A policy concept refers to a specific policy idea or objective, such as constructing a dam, imposing a licensing moratorium, or introducing licence trading. In total, 102 policy concepts were identified and included in the analysis. A policy disposition simply refers to whether an actor is supportive, unsupportive, or neutral about a stated policy concept. The end products of this analysis are tables 4.1, 4.2, and 4.3 (below), which provide concise and graphic summaries of the prevailing policy recommendations in 1978, 1995, and 2002/5, respectively. The tables show the most frequently cited policy concepts at the time as well as various actors' dispositions towards these concepts. By grouping the actors into clusters of similar policy recommendations, ideological groupings were identified.

Shared policy core beliefs are a defining feature of advocacy coalitions, but so is some degree of collective action among the like-minded policy actors. So, in addition to the content analysis of stakeholder submissions, this chapter includes an analysis of the collective action efforts of the identified ideological groupings. This is done primarily by examining the creation and maintenance of umbrella organizations, whose existence provides conclusive evidence of collective action. This is supplemented with anecdotal examples of further collective action efforts, many of them gleaned from stakeholder interviews. All of this evidence is used to determine if and when the ideological groupings became veritable advocacy coalitions, thereby establishing the most important political forces in the Alberta water policy subsystem.

The Era of Aggie Dominance

In 1975, during their first election campaign as the incumbent government, the PCs announced plans for the construction of a dam on the Oldman River. Interest in such a dam dated to at least 1912 and resurfaced every few years, particularly during dry spells. The purpose of an Oldman River dam was to provide additional water storage and greater control of river flows; this would increase the water supply available for downstream irrigators and make the water supply much more reliable. By the mid-1970s, water scarcity was severely constraining further irrigation development in the Oldman sub-basin, and a large new dam was regarded as an effective means of overcoming this scarcity. It also helped that the dam was enormously popular in the irrigation belt, an area where the PCs were hoping to make electoral inroads.

After the PCs gained re-election, a lengthy study and planning period was initiated. In 1978, the Environment Council of Alberta, an independent public body with a mandate "to review government policies and programs and inquire into matters pertaining to environment conservation," reviewed the proposed dam and held open public consultations throughout southern Alberta (Provincial Archives of Alberta 2006, 190). It is this set of public consultations that was included in the content analysis and that provides a snapshot of the actors, policy core beliefs, and advocacy coalitions in the southern Alberta water policy subsystem in the late 1970s. Given the nature and purpose of the consultations, it is not surprising that the most frequently cited policy concepts at this time related to dam construction and irrigation expansion. These are also the policy concepts on which there was most disagreement, providing good indications of the ideological divisions in the subsystem at the time.

As shown in table 4.1, the subsystem in 1978 was ideologically divided between a larger group of Aggies and a smaller group of Greens. This suggests the presence of two advocacy coalitions, but further probing reveals that only one of the ideological groupings – the Aggies – meets the second essential criterion of an advocacy coalition: a non-trivial degree of collective action. Though the Greens shared common policy core beliefs, they were a disparate and uncoordinated movement at this time, disqualifying them as a fully fledged advocacy coalition. This thesis is explored further below, first through an examination of the Aggies and their tightly knit advocacy coalition, and then through an examination of the Greens and their lack of concerted collective action.

Table 4.1 Water Policy Recommendations, 1978

Policy concepts	Expand irrigation	Increase environmental water	Water pricing	Take mitigation/ remediation measures	Increase water use efficiency	Build dams/ storage	Sub-basin equity in water development
AGGIE BELIEFS							
Alberta Irrigation Projects Assoc.	Y			Y	Y	Y	
Eastern ID	Y		N	Y	Y	Y	Y
Lethbridge Northern ID	Y					Y	
Magrath ID	Y			Y		Y	
Taber ID	Y			Y	Y	Y	
St Mary River ID				Y	Y	Y	
Bow River ID	Y						Y
Medicine Hat C of C	Y					Y	
Picture Butte C of C	Y				Y	Y	
Brooks C of C						Y	
Bow Island C of C	Y					Y	
Bow River Gas Co-op	Y				Y	Y	
Unifarm	Y	Y		Y	Y	Y	
Town of Bow Island	Y			Y		Y	
Town of Picture Butte	Y			Y	Y	Y	
Town of Taber				Y	Y	Y	
Village of Foremost	Y			Y		Y	
Pincher Creek MD						Y	
Southeast Alberta Regional Planning Commission	Y			Y			Y

Table 4.1 Water Policy Recommendations, 1978 (Continued)

Policy concepts	Expand irrigation	Increase environmental water	Water pricing	Take mitigation/ remediation measures	Increase water use efficiency	Build dams/ storage	Sub-basin equity in water development
Oldman River Regional Planning Commission	Ø		Y	Y	Y	Y	
Electric Utility Planning Council*						Y	
Jensen, Roy	Y					Y	
Nemeth, Frank	Y				Y	Y	
Schuld, Peter	Y				Y	Y	
GREEN BELIEFS							
Alberta Fish & Game Assoc.*	N			Y	Y	Y/N	
Federation of Alberta Naturalists*	N	Y	Y			Y/N	
Girl Guides of Canada*						N	
Cradduck, Mervin*	N			Y	Y	N	
Pharis, Hilton*						N	
INDETERMINATE							
Town of Pincher Creek				Y	Y	Ø	
Eisenhauer, John	Y			Y	Y	Y/N	

*Not part of an advocacy coalition – no evidence of collective action with like-minded actors
Y = yes/supportive; N = no/unsupportive; Ø = neutral
C of C = chamber of commerce; ID = irrigation district; MD = municipal district

The Aggie coalition comprised a number of different actors, many of which show up in the 1978 data: the irrigation districts, irrigation-reliant businesses, larger agri-food advocacy organizations, irrigation belt municipalities, and regional planning commissions. To this list must be added other actors who did not or could not testify in the consultations, such as irrigation belt MLAs, the Department of Agriculture, and at least part of the recently formed Department of the Environment. Many of these actors had worked together for a long time, so that their collective constitution as an advocacy coalition is in little doubt.

The heart of the Aggie coalition has long been the irrigation districts and their advocacy organization, the Alberta Irrigation Projects Association. Today, the irrigation districts are farmer-run cooperatives, though in the past many have operated as Crown corporations of either the Alberta or federal governments. In the 1978 hearings, the irrigation districts were ubiquitous. Representatives from the Eastern, Lethbridge Northern, Magrath, Taber, St Mary River, and Bow River Districts gave testimony, as did a number of individual irrigators from within the districts, such as Roy Jensen, Frank Nemeth, and Peter Schuld. The Alberta Irrigation Projects Association also testified as a distinctive entity, sending eight of its directors and administrators to make statements at various public hearings. Given that the irrigation districts, particularly Lethbridge Northern, would be the prime beneficiaries of the Oldman Dam, their extensive involvement in the hearings is not surprising.

Other actors, who are not irrigators but benefit from irrigation, have also been important members of the Aggie coalition. They have included rural businesses, from farm equipment suppliers to food processors, whose livelihoods rely in some way on the presence of irrigation. In 1978, these actors were represented mostly by local chambers of commerce, such as those from Brooks, Picture Butte, and Medicine Hat, though a few irrigation-reliant businesses also testified on their own behalf. This category also includes irrigation belt municipalities who benefit from the employment, population, and tax base that irrigation helps to sustain. A number of these municipalities made submissions in favour of the Oldman Dam, including the Town of Bow Island, the Town of Picture Butte, the Town of Taber, the Village of Foremost, and the Municipal District of Pincher Creek.

Larger agri-food advocacy organizations have also been prominent Aggies, since irrigation cross-cuts many of the traditional agricultural sectors. The best example of this from the 1978 hearings was Unifarm, a producer-funded farmer advocacy organization that has since been renamed the Alberta Federation of Agriculture (Alberta Federation of Agriculture 2014).

The Aggie coalition has also long included a number of provincial government organizations that believe in the promise of irrigation and actively promote it. The most evident example from the 1978 hearings was the regional planning commissions, the Southeast Alberta Regional Planning Commission and the Oldman River Regional Planning Commission. The regional planning commissions were provincially created bodies designed to support the planning activities of local governments, and a number of them championed irrigation as a means of economic development in dry and sparsely populated regions. Not evident in the 1978 data was the presence of the Department of Agriculture and at least part of the Department of the Environment, though all other evidence points to them as being important Aggie coalition members. The Department of Agriculture has long been a tireless advocate of irrigation within the Alberta government, and irrigators and departmental administrators have shared a close working relationship for decades. The same can be said of the Water Resources Division in the Department of the Environment, the division responsible for dam construction and operation, and transferred to the Environment department (from Agriculture) when the department was newly formed in 1971. These departments were dominated by officials with training in irrigation-friendly fields, such as agronomy and engineering, and also benefited professionally from the introduction and continuation of irrigation-related government programs. Officials from the Agriculture and Environment Departments were precluded from testifying in the 1978 hearings but were actively involved in the policy subsystem and were prominent members of the Aggie coalition.

Rounding out the Aggie coalition were the irrigation belt MLAs, also not evident in the 1978 hearings but key to the Aggie coalition and its policy aspirations. Most irrigation belt MLAs, some of whom were farmers or irrigators themselves, are true believers in irrigation in the sense that they believe in irrigation's potential as a means of rural economic development. They are also beneficiaries of irrigation, as they are electorally rewarded for securing new irrigation and water management projects, or preserving existing ones, for their constituents. Starting with the 1975 election, the irrigation belt became a stalwart part of the PC voting base, consistently returning PC MLAs to the legislature and creating something of a symbiotic relationship between irrigators and the perennially governing party. This relationship continued until the 2012 election, when the irrigation belt shifted en masse to the upstart Wildrose Party.

Despite its disparate membership, the Aggie coalition was united in the belief that continuous irrigation expansion is the path to prosperity in southern Alberta. Irrigation allowed for intensive farming and the growth of cash crops where otherwise there would be only dryland ranching. In this way, irrigation supported relatively high rural population densities, including small-scale family farms and rural towns and cities. Irrigation was an economic activity in which irrigators clearly had a material self-interest, but it was also a pillar of rural communities supporting a distinctive, idyllic way of life. Aggies further believed that irrigation expansion should be supported by an extensive, state-funded water management infrastructure. They believed they were entitled to a water management infrastructure in the same way that other industries were entitled to the public road and rail infrastructure that supported them. Water shortages and environmental problems, to the extent they were recognized, could be managed through further dam construction and the inclusion of remediation and mitigation projects, such as fish ladders and recreational areas. In short, the Aggies' approach to water governance was a decidedly hard one, based on maximum use of available water resources and a supply-side approach to attaining maximum use.

Collective action and policy advocacy among irrigation enthusiasts dates back to the late nineteenth century, giving the Aggie coalition a remarkably long pedigree in water policymaking. In 1894, an entity known as the South-Western Irrigation League of the North-West Territories was formed during an irrigation convention in Calgary and sent a delegation to Ottawa, urging the government to pass the Northwest Irrigation Act and to begin hydrographic surveys of the Prairies (Mitchner 1973, 70). The league comprised sitting senators and MPs from Alberta, sitting members of the Territorial Legislature, local mayors, and local agricultural societies and boards of trade (de Loë 1994, 68; Lee 1966). In 1907, western farmers, businesses, government agencies, and railway companies with an interest in irrigation formed the Western Canada Irrigation Association, a more enduring advocacy organization that spanned all four western provinces. Through its lobbying efforts, the association was instrumental in the passage of the Alberta Irrigation Districts Act and in securing millions of dollars in provincial support for irrigation projects. Despite these successes, the logistical challenges of maintaining collective action across the vast expanse of western Canada proved too much, and the association was disbanded in 1926 (Mitchner 1973, 286–9).

By the late 1970s, collective action in the Aggie coalition was based on three coordinative bodies: the Alberta Irrigation Projects Association, the Irrigation Council, and the irrigation caucus. First incorporated in 1946, but operating less formally prior to that, the Alberta Irrigation Projects Association picked up where the Western Canada Irrigation Association left off. It served as the umbrella organization for the irrigation districts, coordinating their policy advocacy activities and forging close working relationships with government legislators and administrators. These relationships were solidified even further through the Irrigation Council, a provincial body first created in 1920 with a mandate to oversee the operations of the irrigation districts. Members of the council were appointed by the minister of agriculture and involved a mix of government officials and irrigators. During the 1970s, the council had the additional responsibility of administering the province's Irrigation Rehabilitation Program, putting substantial resources at its disposal (Provincial Archives of Alberta 2006, 68–9). Within the governing PC Party, there was also an irrigation caucus involving PC MLAs elected from the irrigation belt and dealing exclusively with irrigation issues. The irrigation caucus routinely consulted with the Alberta Irrigation Projects Association, Alberta Agriculture officials, and other members of the Aggie coalition, and allowed irrigation belt MLAs to coordinate strategies for advancing irrigation interests within the government.

In sum, the Aggies of the late 1970s had all the characteristics of a textbook advocacy coalition. The results of the content analysis clearly show a grouping of pro-irrigation actors that, in combination with a few key actors precluded from testifying in the public consultations, constituted the main membership of the Aggie coalition. The further fact that these actors had umbrella organizations and other coordinating bodies confirms that they were not just ideological compatriots, but a fully fledged advocacy coalition. This Aggie coalition has been of central importance in the development of water policy in southern Alberta, as will be shown in chapters 6, 7, and 8.

In addition to the Aggies, the content analysis also showed a Green ideological grouping. The Greens shared a common belief that further irrigation expansion was undesirable, given the deteriorating state of southern Alberta's riverine environments. They believed that environmental protection should be the first priority of Alberta's water policy and that irrigation should be allowed only to the extent that it was not environmentally harmful. It is interesting to note, however, that even the Greens' policy core beliefs were heavily influenced by the hard approach to water governance that was dominant at the time. The two main Green groups in the 1978 hearings, the Alberta Fish and Game Association and the Federation of Alberta Naturalists, both testified in support of constructing new off-stream water storage projects as an alternative to the construction of new on-stream dams, judging off-stream storage to be less environmentally damaging than on-stream storage, but accepting that some new storage projects were inevitable.

This was how deeply rooted the hard approach to water policy was in the late 1970s.

The Greens were a much smaller ideological grouping than the Aggies and comprised two types of actors: conservationist groups and environmentalist groups. The main conservationist group was the Alberta Fish and Game Association, which has existed in various forms since 1908. Its membership is made up mostly of recreational fishers and anglers who rely on healthy rivers to support their pastime (Alberta Fish and Game Association 2014). Some of them are also farmers and most tend to have a more instrumental view of nature than the environmentalists, who tend to value nature in and of itself. In the late 1970s, the only major environmentalist group in the water policy subsystem was the Federation of Alberta Naturalists, which was created in 1970 through the alliance of six natural history clubs. The federation was dedicated to the preservation and study of nature and gradually became involved in policy advocacy towards these ends (Nature Alberta 2013).

Beyond the Alberta Fish and Game Association and the Federation of Alberta Naturalists, there were few other unambiguously committed Greens. The only other organization showing Green beliefs in the policy subsystem was the Girl Guides of Canada, whose presence is curious, given that, other than their testimony in public hearings, they showed little evidence of policy advocacy activities. There were also a couple of individuals named Mervin Cradduck and Hilton Pharis who show up as Greens in the 1978 data, but neither could be characterized as a committed environmentalist. Cradduck was a dryland farmer who was more anti-irrigation than pro-environment, because he felt irrigation was a threat to dryland farming (Cradduck 1978). Pharis had been a member of a special committee that, earlier in the year, had produced feasibility studies recommending the dam, and he testified to express his dissent with that recommendation (Pharis 1978). There is also little evidence of committed Greens in the ranks of Alberta's legislators and administrators. Alberta has never elected a Green Party MLA and, though there were probably some Green sympathizers in the Department of the Environment at that time, the Environment department, as an organization, has never aligned itself closely with environmentalist groups, for reasons explained further below.

So, in the late 1970s, the Greens were really nothing more than a handful of advocacy groups with similar policy core beliefs, but little collective action among them. It was not until the mid-1980s that the Alberta Environment Network and the Friends of the Oldman River would be formed, the two umbrella organizations that would come to define the Greens as a fully fledged advocacy coalition. Until then, some Greens were present in the policy subsystem but working largely on their own towards distinctive conservationist and environmentalist ends.

This left the Aggies as the dominant coalition in the subsystem, and their hard approach to water governance was so deeply engrained that even the major Green groups acceded to some hard policy ideas. Borrowing from analyses of water policy processes in the American West, Glenn (1999) describes Alberta water policymaking as dominated by an "iron triangle" of irrigation proponents, the three sides of the triangle being irrigators, irrigation bureaucrats, and irrigation belt MLAs. The evidence from this study confirms this as an apt description, at least until the late 1980s and early 1990s, when the Greens began to coalesce into a concerted political force.

The Emergence of Aggie-Green Pluralism

The decision to construct the Oldman Dam was both the zenith of the hard policy era and a turning point in Alberta's water politics. Opposition to the dam served as a rallying point for environmentalists and conservationists who came together into a Green advocacy coalition for the first time. It also prompted much closer scrutiny of Alberta's water allocation and governance policies, the main principles of which had not been reformed since the 1894 Northwest Irrigation Act. In 1990, at the height of the Oldman Dam conflict, senior officials advised Cabinet that a review of Alberta's water legislation was necessary. A review was then initiated in 1991, eventually resulting in the drafting of new water legislation in 1994. The draft legislation served as the focal point for extensive public consultations in 1995, consultations that involved a larger and more diverse array of actors than in 1978. It is these 1995 consultations that provided the raw data that were included in the content analysis.

While the 1978 consultations had been narrowly focused on the merits of the Oldman Dam, the 1995 consultations were more of a "blue sky" exercise in which all aspects of Alberta water policy were thrown into question. This resulted in a much larger number of policy concepts entering the discussion, only some of which eventually found their way into government policy. A look at the policy concepts included in table 4.2 also shows that the policy discussion was much softer in 1995 than it had been in 1978. Softer concepts such as water licence trading, water pricing, environmental water recovery, licensing moratoria, minimum streamflows, and many others were quite prominent in 1995, compared to 1978 when the discussions were mostly about dam construction and irrigation expansion. This represents a major shift in the water policy debate.

Table 4.2 summarizes the results of the content analysis from 1995, listing the relevant actors and their policy recommendations. The evidence shows the presence of Aggie and Green ideological groupings, with some notable membership changes in each, compared to 1978. Moreover, analysis of their collective action efforts reveals that both groupings qualified as bona fide advocacy coalitions, transforming the water policy subsystem from a single-coalition-dominant subsystem to a two-coalition-pluralistic subsystem with genuine ideological competition in the development of water policy.

The Aggie coalition did not undergo major change between the late 1970s and mid-1990s, though it did experience some marginal changes in membership. Most of the major Aggie players from 1978 also appear in the 1995 data, including the irrigation districts (including the Alberta Irrigation Projects Association), the larger agri-food sector (Unifarm, and the Alberta Cattle Commission), and the irrigation belt municipalities (such as Rocky View, Newell, and Vauxhall). The Department of Agriculture and irrigation belt MLAs also remained steadfast partners in the Aggie coalition, though their positions in government precluded them from participating in the public consultation process. The regional planning commissions no longer show up in the data because they had been dismantled during Premier Klein's government downsizing and restructuring. Irrigation-reliant businesses also do not appear in the 1995 or the 2002/5 data, suggesting that their participation in the Aggie coalition faded over time, for some reason. It is also interesting to note the indeterminate and somewhat tortured position of the City of Medicine Hat, which was situated in the irrigation belt but had a vocal environmentalist constituency and seemed caught between these two realities. In short, it seems that some members of the Aggie coalition had faded in the 1990s, but the coalition's integrity remained intact.

Two interesting actors who show up in the 1995 data and who appear to have Aggie policy core beliefs, but are not really Aggies, are the Canadian Association of Petroleum Producers and the TransAlta Corporation. The Canadian Association of Petroleum Producers represented the oil industry in Alberta, which had an interest in water policy, since water is an essential component of oil extraction. TransAlta was a private utility company that operated a number of hydroelectric dams in the upper reaches of the Bow sub-basin and had an obvious interest in preserving them. Both actors involved themselves in the development of the Water Act, primarily to ensure that there were no radical changes to the policy status quo that might threaten their operations, apart from the introduction of water licence trading, which both supported. This belief in preserving the status quo gave them overlapping beliefs with the Aggies, but there is no evidence that they engaged in substantive collective action with them. Nor is there any evidence that they engaged in collective action with each other to form their own distinctive advocacy coalition. Accordingly, the oil industry and TransAlta are best characterized as arm's-length Aggie supporters, sharing many of the coalition's beliefs and endorsing many of its policy positions, but not actively participating in the coalition itself.

The most significant changes for the Aggie coalition between the late 1970s and mid-1990s were in their coordinative bodies. The Alberta Irrigation Projects Association and the Irrigation Council both remained active and important arenas of collective action, and to these can be added the Alberta Cattle Commission and the Southern Alberta Water Management Committee. The Alberta Cattle Commission was formed in 1969 as an umbrella organization for Alberta's beef industry and became involved in water policy in response to the stock-watering concerns of beef producers and the considerable overlap between beef producers and irrigators. The commission's presence in 1995 was notable not only because of its submission during the public consultation process, but also because it mobilized and coordinated Aggie opposition to the legislation eventually proposed, which we will examine more closely in chapter 7. The Southern Alberta Water Management Committee was formed in 1987 as an umbrella organization to coordinate the advocacy efforts of the pro-dam side in the Oldman Dam conflict. It played a key role in the Aggies' collective action efforts of the late 1980s and early 1990s, but was disbanded after the dam was built. It was also in the early 1990s that the irrigation caucus of the governing PC Party faded away. The circumstances around its demise are unclear, but it was a substantial blow to the Aggie coalition that is lamented by irrigators even today. The irrigation belt MLAs remained as members of the coalition, but their interactions with irrigators were no longer as routinized as they once were.

In sum, the Aggies experienced marginal changes in membership and more substantial changes in collective action, but they remained

Table 4.2 Water Policy Recommendations, 1995

Policy concepts	Preserve status quo prioritization	More environmental water	Water licence trading	Water pricing	Crown reservations	Environmental water recovery without compensation	Environmental water recovery with compensation	Private conservancy	Licensing terms/ conditions	Licensing moratoria	Minimum streamflows	Greater regulation of water use
AGGIE BELIEFS												
Alberta Irrigation Projects Assoc.	Y		Y	N			Y		N	Y		
Bow River ID	Y		Y	N			Y			N		
Eastern ID	Y		Y	∅							∅	
Lethbridge Northern ID	Y			N					N		∅	
Unifarm			∅	N		N						
Alberta Cattle Commission	Y		Y	Y		N			N			
TransAlta Utilities Corp.*			Y	∅			Y		N			
Canadian Assoc. of Petroleum Producers*	Y		Y	N								
County of Newell	Y		Y	N		∅						Y
MD of Rocky View	Y		∅	N					N			
Town of Vauxhall	Y		Y	N			Y			N		
GREEN BELIEFS												
Alberta Fish & Game Assoc.	N		N	∅	Y	Y	Y	Y	Y			Y
Environmental Law Centre	N		∅	∅	Y	Y	Y	Y				
Friends of the Oldman River	N	Y		Y		Y			Y		Y	Y
Red Deer River Naturalists	N	Y	N	Y					Y			
Southern Alberta Environmental Group	N	Y	N	Y					Y			Y

Grassland Naturalists			N		Y				Y	Y	Y	Y
Girl Guides of Canada*			N	Y		Y						
Southern Alberta Development and Protective Assoc.			N							Y		
Trout Unlimited	N	Y	Y			Y	Y	Y	Y		Y	
Bow River Water Quality Council			Y				Ø		Y	Y		
Little Red River Cree Nation*	N	Y	N	N	Ø							
MD of Clearwater*			N	Ø					Y			Y
Dixon, J.S.	N			Y						Y		
Bradley, Cheryl				Y		Y		Y	Y		Y	
Bankes, Nigel	N		Y	Y								
Kure, Elmer			N	Y		Y						
INDETERMINATE												
Alberta Golf Industry Assoc.				N								
Ducks Unlimited			Y									Y
Municipal District of Foothills				N		Y						
Rew, Doreen			Y	Y		N			Y			
Special Areas Board			Y	N		Y			Y			
Bow Waters Canoe Club	Y/N								Y			
City of Medicine Hat	Y		Ø		Y	Y						
Town of Cochrane	Y		Y		Y	Y/N			Y	Ø		
Calgary C of C	N		Y	Y	Y	N	Y	Y	N			Y

*Not part of an advocacy coalition – no evidence of collective action with like-minded actors
Y = yes/supportive; N = no/unsupportive; Ø = neutral
C of C = chamber of commerce; ID = irrigation district; MD = municipal district

a sizeable and tightly knit advocacy coalition through the mid-1990s. Nevertheless, they no longer had complete domination over the water policy subsystem, as the Greens expanded their presence and gradually morphed into a fully fledged advocacy coalition in the mid-to late 1980s. This was indeed a major change to a subsystem that had featured only a single advocacy coalition for its entire, decades-long existence.

In the mid-1990s, the Green coalition was still composed primarily of conservationist and environmentalist groups, but there were now many more of them. The original conservationist group, the Alberta Fish and Game Association, remained a steadfast Green participant, but they were now joined by other conservationist groups such as Trout Unlimited and Ducks Unlimited. Trout Unlimited had formed in Canada in 1984 and, by the mid-1990s had eleven chapters in Alberta, all dedicated to the preservation of trout stocks and habitat (Trout Unlimited 1995). Ducks Unlimited shows up in the 1995 content analysis as indeterminate, but the 2002/5 data, interview data, and their organizational mandate mark them as Greens.

The conservationist groups were important but the main growth in the Green coalition came from the proliferation of environmentalist groups, some of which were recently formed and others that had existed for awhile but were new subsystem participants. The early naturalist groups evident in 1978 carried over into 1995 with both the Red Deer River Naturalists and the Grasslands Naturalists making submissions during the 1995 public consultations. The naturalists were now joined by newer participants such as the Alberta Wilderness Association,[5] the Southern Alberta Group for the Environment, and the Environmental Law Centre, each of which was a substantial addition to the Green coalition.

The Alberta Wilderness Association was founded in 1965 in rural southwestern Alberta but eventually grew into a province-wide organization "dedicated to the completion of a protected areas network and the conservation of wilderness throughout the province" (Alberta Wilderness Association 2014). They became involved in the water policy subsystem in the mid-1980s, primarily in objection to the Oldman Dam.

5 The Alberta Wilderness Association does not show up in the 1995 data, but it made submissions in 1984 and 2005 and was part of the Friends of the Oldman River in the interim, so it probably deserves inclusion in the Green coalition of the mid-1990s.

The Southern Alberta Group for the Environment was a much newer organization, formed in 1984 by a group of dissidents from the Lethbridge Fish and Game Association. When the association's board decided to support the Oldman Dam, a group of opponents, led by an environmentalist named Martha Kostuch, left the association to form their own group more firmly dedicated to environmental protection. Since its inception, the Southern Alberta Group for the Environment has maintained a strong presence in the policy subsystem and has produced some of the Greens' strongest leaders, including Kostuch. The Environmental Law Centre was formed in 1982 by a group of environmentalist law faculty at the University of Calgary. "The Centre's overall objective is making the law work to protect the environment," and it has pursued this goal by critiquing policy and helping to craft Green alternatives (Environmental Law Centre 1995, 3). These new actors not only helped to swell the Greens' ranks, they also brought with them new skills and resources to advance the Green cause.

The 1995 data also show a number of individuals with Green beliefs, most of which were prominent leaders of one of the Green groups described above. Elmer Kure was a well-respected leader in the Alberta Fish and Game Association, Cheryl Bradley was a top-ranking member of the Southern Alberta Group for the Environment, and Nigel Bankes was a driving force behind the Environmental Law Centre. J.S. Dixon appears to have been an active citizen with environmentalist concerns, though not a Green group leader. These individuals are noteworthy because their commitment and tireless work was very important in advancing the Green cause and in holding it together as a political movement.

There were also two categories of actors in the 1995 data who held Green beliefs but were not really Green coalition members. They could be characterized as arm's-length Green supporters in the same way that the Canadian Association of Petroleum Producers and TransAlta were arm's-length Aggie supporters.

One of these categories was the Red Deer sub-basin municipalities. Most of these municipalities wanted to prevent the same kind of overallocation in the Red Deer as had occurred in the three southern sub-basins, and they also wanted to prevent further water development in the southern sub-basins that would require more Red Deer water in order to satisfy Alberta's apportionment obligation to Saskatchewan. This gave them policy core beliefs that overlapped with the Greens', but there is little evidence of collective action between them.

The other category of actors was First Nations. Greens and First Nations share many policy core beliefs in protecting and restoring riverine environments, but First Nations have water rights claims and cultural attachments to water that set them apart from Greens. First Nations have generally been reluctant to engage in the regular policy consultation processes in which Greens have been very active, preferring instead to deal directly with the Alberta government on a nation-to-nation basis. There was also a split between Greens and First Nations in their resistance to the Oldman Dam when a small group of Natives attempted to resist the dam's construction through use of force, a tactic eschewed by the Green groups. Overall, though the policy positions of the First Nations typically lined up with the Greens, it would be a stretch to say that they were Green coalition members, and instead can be characterized as arm's-length Green supporters.

One of the biggest differences between the Greens of 1978 and those of 1995 was in their collective action. In the 1980s, two coordinative bodies emerged that vastly expanded and improved Green collective action efforts: the Alberta Environment Network and Friends of the Oldman River.

The Alberta Environment Network was formed in 1980 to facilitate cooperation among government agencies, businesses, professional organizations, and advocacy groups interested in the environment. The network itself has not been an advocacy group and has not taken up environmental causes; instead, it has built and maintained a directory of environmental organizations, published a bimonthly newsletter, and helped coordinate campaigns on various environmental issues through its issue-based caucuses (Stefanick 1991, 88–91). Many of the Greens interviewed for this study indicated that, through the network and its water caucus, they are regularly in contact with each other – at least monthly – and they share information and coordinate strategy on issues of common interest. In short, the network provides the essentials needed for collective action in an advocacy coalition.

Even more important to the collective action efforts of the Greens was Friends of the Oldman River. This organization was created in the summer of 1987 as a coalition of Green organizations opposed to the construction of the Oldman Dam. Friends of the Oldman River was important because it brought together all Greens with an interest in water issues into a serious political campaign involving protests, public education, fundraising, lobbying, and court action. Though the group faded away in the late 1990s as the Oldman Dam issue became settled, its legacy of relationship-building and political experience cannot be overstated. If one had to pinpoint the exact moment that a fully fledged Green advocacy coalition emerged in the water policy subsystem, it may well have been that summer day in 1987 when Friends of the Oldman River was formed.

Conspicuous by its absence from the Green coalition was the Department of the Environment. Since the Department of Agriculture has long been a stalwart of the Aggie coalition, one might expect that the Department of the Environment would logically fit into the Green coalition. After all, it has an environmental protection mandate, and individuals with environmentalist sympathies are likely to be attracted to the organization. While certainly some individuals in Alberta Environment harbour Green beliefs – and some have worked on behalf of Green causes – the organization, as a whole, is much closer to a policy broker than a Green coalition member.

As described in chapter 2, policy brokers are actors who position themselves between advocacy coalitions in a policy subsystem, often attempting to bridge the differences between them. From its inception, the Department of the Environment has always been something of a policy broker, since it has housed both dam-builders (with Aggie sympathies) and environmental regulators (with Green sympathies). Its brokerage position became even more pronounced in the 1990s in the wake of the Oldman Dam controversy and under the ministerial leadership of Ralph Klein. As a new environment minister, Klein was eager to mend the rift between Aggies and Greens caused by the Oldman Dam, and, as an effective practitioner of brokerage-style politics, he positioned Alberta Environment as an honest broker between the two coalitions, open to views from both. After Klein moved on to the premiership, the department maintained its brokerage role and continues so today.

Overall, the content analysis data show that, by 1995, the Alberta water policy subsystem had evolved from a single-coalition-dominant subsystem into a two-coalition-pluralistic subsystem characterized by ideological competition between Aggie and Green coalitions. In addition, a policy broker had emerged in the form of Alberta Environment, creating something of a bridge between the two coalitions. The shift from one coalition to two coalitions occurred in the mid- to late 1980s as the Greens expanded and improved their collective action efforts, while the policy broker emerged in the early 1990s as Environment Minister Ralph Klein repositioned the department in the wake of the Oldman Dam conflict. As we will see in chapter 7, the emergence of the Greens as an advocacy coalition and the Environment Department as a policy broker played major parts in Alberta's shift towards softer water policy, which started with the Water Act reforms of 1996.

The Normalization of Aggie-Green Pluralism

By the early 2000s, the Water Act was on the books and focus had shifted to its implementation. The Water Act introduced a number of new (softer) water policy instruments, including water licence trading, conservation holdbacks, and water conservation objectives. It also contained provisions for new water management planning through which precise rules for water licence trading and conservation holdbacks and specific levels for water conservation objectives could be defined on a basin-by-basin basis. In effect, the Water Act planted the seeds for another round of water policy development and, because the SSRB was the most water-stressed basin in the province, its water management planning process was undertaken first. For reasons explained further in chapter 8, Alberta Environment split the planning process into two phases, the first one dealing with water trading and related issues in 2001–2, and the second one dealing with environmental flows and related issues in 2003–5. Since both phases were part of the same planning process, the data from 2002/5 were treated as two parts of a single policy process for the purposes of the content analysis.

The data from 2002/5 are also not quite as good as the data from the earlier policy processes, mostly as the result of the nature of the planning process itself. Development of the SSRB Water Management Plan was very much an exercise in co-management, with stakeholders from both coalitions actively involved. Basin advisory committees were established for each of the SSRB's sub-basins, and each basin advisory committee comprised a wide cross-section of stakeholders from that sub-basin, with both the Aggie and Green coalitions well represented. The basin advisory committees were also involved from start to (almost) finish in developing the SSRB Water Management Plan, ultimately making recommendations to Alberta Environment on what should be included in the final plan. Both phases featured open public consultation processes, but, with so many of the stakeholders already involved on the basin advisory committees, fewer felt the need to make public submissions. Accordingly, the public submissions

Table 4.3 Water Policy Recommendations, 2002/5

Policy concepts	Preserve status quo prioritization	More environmental water	Expand irrigation	Water licence trading	Environmental water recovery without compensation	Build dams/ storage	Operate dams to restore rivers/ wetlands	Greater regulation of water use	Crown reservations	Licensing moratoria	Private conservancy	Sub-basin equity in water development
AGGIE BELIEFS												
Bow River ID		N								N		
Eastern ID	Y				∅							
Lethbridge Northern ID					∅							
St Mary River ID	Y					Y	∅		N			
Alberta Beef Producers				Y								
Special Areas Board			Y			Y						
GREEN BELIEFS												
Alberta Fish and Game Assoc.		Y			Y					Y		
Alberta Wilderness Assoc.	N	Y								Y		
Trout Unlimited		Y			Y					Y	Y	
Environmental Law Centre			N	Y	Y			Y	Y	Y	Y	
Ducks Unlimited	N	Y										
Bow River Basin Council		∅			Y			∅		Y		
City of Red Deer*		Y	N	Y	Y		Y		Y	Y	Y	Y
Clearwater County*		Y		Y	Y		Y		Y			Y
County of Stettler*		∅					Y			Y		Y
Bankes, Nigel, & Kwasniak, Arlene		Y			Y		Y				Y	
Kure, Elmer				N				Y				
Rew, Doreen*			N					Y		Y		
INDETERMINATE												
Starland County		Y		Y	Y	Y	Y	Y		∅		Y

*Not part of an advocacy coalition – no evidence of collective action with like-minded actors
Y = yes/supportive; N = no/unsupportive; ∅ = neutral
ID = irrigation district

very likely underrepresent the actors who were actively involved in the 2002/5 policy process. Efforts were made to compensate for this,[6] but we must be somewhat more circumspect in drawing conclusions from these data than the earlier data.

As far as the advocacy coalitions are concerned, the content analysis data from 2002/5 suggest that there was little change from 1995. The Aggie coalition appears much smaller than it was, but this has more to do with the shortcomings of the data than anything else. An examination of the lists of participants on the basin advisory committees shows that representatives from the irrigation districts (including the Alberta Irrigation Projects Association), irrigation-reliant businesses, the larger agri-food sector, and the irrigation belt municipalities were all involved, even though many of them did not make public submissions. The Department of Agriculture continued its close relationship with irrigators, and irrigation belt MLAs continued to support and represent irrigation causes. The Green coalition continued to comprise a combination of conservationist and environmentalist advocacy groups, many of which had representatives on the basin advisory committees and also made public submissions. The Red Deer sub-basin municipalities were also well-represented and continued as arm's-length Green supporters. In short, it seems that the advocacy coalitions had reached a point of equilibrium between 1995 and 2005, with few notable membership changes on either side.

The same can be said of their collective action efforts. Neither coalition created or disbanded any coordinative mechanisms during this period, so that their collective action efforts in 2002/5 remained much as they were in 1995. Since these collective action efforts were substantive and non-trivial, both the Aggies and the Greens easily qualified as advocacy coalitions in the development of the SSRB Water Management Plan.

One important thing that turns up in the 2002/5 data is preliminary evidence of policy-oriented learning on two key policy concepts. The 1995 data showed the Greens to be steadfastly against water licence trading and the Aggies to be strongly against uncompensated environmental water recoveries. However, by 2002/5 these beliefs seem

6 One method of compensation was to consider all members of the basin advisory committees as active participants in the policy subsystem in 2002/5 for the purpose of sampling actors, even though public submissions do not exist for most of these actors. Another method of compensation was interview data from key process participants.

to have softened and, in some cases, reversed. Most Greens withdrew their previous objections to water licence trading, neither supporting nor opposing it, and some even reversed their previous positions to support water trading. Similarly, most Aggies withdrew their opposition to uncompensated water recoveries, and some even expressed their neutrality on the issue. These belief changes are quite important, suggesting that policy-oriented learning was taking place in both advocacy coalitions, a hypothesis that will be followed up further in chapter 8 when we examine the formation of the SSRB Water Management Plan.

Overall, the chief difference between 1995 and 2002/05 was the normalization of the two-coalition-pluralistic structure that first emerged in the late 1980s. Even into the mid-1990s, the Greens were regarded by most Aggies and by some government decision-makers as an upstart group of rabble-rousers who were disrupting Alberta's "normal" water politics. Many policy elites expected that the Greens would fade away and that the policy subsystem would soon return to normal. By the early 2000s, these expectations were gone and both the Aggies and government decision-makers had (reluctantly) accepted the Greens as a permanent fixture in the policy subsystem. This is evident in the co-management design of the basin advisory committees and the extensive representation of both coalitions on these committees. It is also evident in both coalitions' increasing acceptance of the need for cross-coalition policy compromises, one of the dominant features of the 2002/5 SSRB water planning process, which will be investigated further in chapter 8.

Explaining the Shift to Softer Policy: The First Piece of the Puzzle

Through a qualitative content analysis of policy elites' submissions during public consultation processes, this chapter has investigated the policy core beliefs prevalent in the southern Alberta water policy subsystem between 1978 and 2005. These data on policy core beliefs, combined with evidence pertaining to actors' collective action efforts, have allowed us to identify the advocacy coalitions active in the policy subsystem at three key points during the period of study. It shows how the subsystem evolved from a single-coalition-dominant structure characterized by Aggie dominance to a two-coalition-pluralistic structure featuring genuine ideological competition between the Aggies and the Greens. The key turning point, in this regard, was the Oldman Dam conflict of the late 1980s and early 1990s, during which the Greens coalesced from a loose ideological grouping into a full-fledged advocacy coalition to challenge the Aggies. By the early 2000s, this Aggie-Green pluralism had become normalized in the policy subsystem, and anecdotal evidence indicates that this subsystem structure remains in place even today.

Noticeably absent from the data were southern Alberta's urban, suburban, and industrial water users. Of the 974 policy recommendations identified and coded in the content analysis, only two made reference to providing more water for urban/suburban use, and only one was unambiguously supportive. Moreover, the City of Calgary was not a recurring participant in the policy subsystem, making a submission only in the 1995 policy consultation process. This absence is curious but was explained by interviewees as stemming from the fact that the City of Calgary already holds large, senior water licences, giving them relative water security. Most of the remaining urban centres in southern Alberta are located in the heart of the irrigation belt, and some even rely on the irrigation districts' distribution systems to obtain some or all of their municipal water supply, so it is not surprising that their beliefs aligned closely with the irrigators'. Also largely absent were the region's industrial water users. The Canadian Association of Petroleum Producers and TransAlta were the only recurring industrial participants in the policy subsystem, but their beliefs were quite close to those of the irrigators, and there is no evidence that they formed a distinctive advocacy coalition of their own. The absence of urban, suburban, and industrial water users sets southern Alberta apart from other water policy subsystems, such as in California, where cities and industries are major water policy players (Munro 1993).

Identifying the advocacy coalitions in the subsystem provides an essential piece of an ACF explanation of Alberta's shift to softer water policy, but only a piece. On the basis of empirical evidence in this chapter, we know that the Aggies were strong proponents of the hard approach to water governance while the Greens believed in a much softer approach. We also know that the Greens emerged as an advocacy coalition in the late 1980s, just a few years before Alberta made its decisive shift towards softer water policy. This suggests a causal linkage between the emergence of the Greens and the shift to softer policy. At this point, however, the case for this linkage is largely circumstantial: we have a new advocacy coalition and a policy change in the direction of the new coalition's policy core beliefs, but we do not know much about how – or even if – the coalition managed to effect this policy change. In the next chapter, we undertake a detailed examination of coalition power resources in the Alberta water policy subsystem to determine whether the Green coalition did indeed have enough power to initiate the shift to softer water policy.

Chapter Five

Coalition Power Resources

It is a truism that almost any sect, cult, or religion will legislate its creed into law if it acquires the political power to do so.

Robert A. Heinlein

The first major policy issue on which the Aggies and Greens clashed was the construction of the Oldman Dam. This project had been on the Aggies' wish list for decades and they viewed it as essential to the prosperity of irrigated agriculture in southern Alberta. For the Greens, the Oldman Dam was one project too many, a guarantee that the already stressed and declining rivers of the region would decline even further in the coming years. While the fight over the dam was clearly a clash of beliefs, it was also a clash of raw political power, a struggle to gain influence over the policy decision-makers who would ultimately determine the dam's fate.

The Greens fought against the dam like a protest movement, using the courts to try to block the dam and cultivating public pressure to try to persuade policymakers to abandon it. Friends of the Oldman River undertook litigation at both the provincial and federal levels, even going all the way to the Supreme Court of Canada to force Ottawa to apply its environmental assessment process to the provincially licenced project. They also organized public protests, circulated petitions, undertook newspaper letter-writing campaigns, and lobbied any government official who would listen. Some of these efforts were quite impressive: a 1989 protest attracted around 800 people, a 1990 petition gained 12,000 signatures, and newspaper coverage of the dam in the major Edmonton and Calgary papers was more negative than positive (de Loë 1999).

The Aggies engaged the Oldman Dam struggle very differently. They were the established political players in Alberta water politics and were very well connected within the Alberta government. The Department of Agriculture was clearly on their side, as were the twelve MLAs elected from the irrigation belt, most of whom were sitting in the governing party caucus, and a couple of whom were sitting in Cabinet. Irrigators also pointed to the important role of irrigation in the southern Alberta economy, arguments that resonated well with vote-conscious policy decision-makers. In the end, the formidable power of the Aggies won the day and they were able to see the dam project through to completion.

The struggle over the Oldman Dam illustrates not only the importance of power in policy processes, but also the diverse sources of power that advocacy coalitions may draw upon. While the Greens relied heavily on protest, persuasion, and public opinion as their sources of power, the Aggies relied on political insiders and economic influence as their sources of power, a pattern that would be repeated many times in the southern Alberta water policy subsystem. The Oldman Dam example also suggests that all sources of coalition power are not equal: the Aggies' influence in the inner halls of policy decision-making overcame the Greens' attempts to persuade and pressure policy decision-makers from the outside.

In this chapter, we will systematically examine the role of power and coalition resources in the shift to softer water policy in southern Alberta. We begin with a brief examination of the role of power in the ACF, the coalition resources recognized in the ACF, and the crucial distinction between hard and soft power. The rest of the chapter then examines the hard and soft power resources of the Aggie and Green coalitions during the period of study, which is essential to investigating the second policy change hypothesis, as outlined in chapter 2.

The ACF and Coalition Power

While considerable ACF research has been dedicated to investigating the role of policy-oriented beliefs in shaping policy outcomes, less effort has been directed towards to the role of power and power resources. Yet "the ACF assumes that access to and exploitation of various political resources is important for advocacy coalitions as they seek to influence public policy" (Jenkins-Smith et al. 2014, 205). In fact, from its inception, the ACF has included a "coalition resources" variable that is meant to encapsulate the relative power of advocacy coalitions and is a clear indication that Sabatier (1993) recognized the importance of power and meant for it to be a part of ACF explanations of policy development. It is only recently, however, that ACF scholars have directly addressed the role of coalition resources, and Jenkins-Smith et al. (2014) identify coalition resources as one of the most important frontiers in ACF research.

Much of the work on coalition resources has thus far focused on developing a typology of resources to allow ACF scholars to understand what constitutes a coalition resource and how exactly it confers power and influence to a coalition. As described in chapter 2, the list of coalition resources identified in the ACF literature currently includes formal legal authority to make policy decisions, financial resources, public opinion, information, mobilizable troops, and skilful leadership (Sabatier and Weible "Advocacy Coalition Framework" 2007, 201–3). It was further argued that a good way to understand the potency of different coalition resources is through Nye's distinction between hard and soft power. Hard power is command-based or coercive power, the ability to push others to do things they might not otherwise do of their own volition. In domestic policy subsystems, hard power typically takes the form of formal legal authority to make policy decisions and is the most potent resource a coalition can have at its disposal. Soft power is more co-optive than coercive and is based on the ability to pull or attract people towards a desired action. Soft power resources tend to be more diverse and ephemeral than hard power resources and, in a domestic policy subsystem, would include things like public opinion, information, mobilizable troops, and skilful leadership. Financial resources can be both a hard or soft power resource, depending on how they are used, but in domestic policy subsystems they are more easily converted into soft power.

Using the distinction between hard and soft power, the second ACF policy change hypothesis was reformulated in chapter 2 to provide a more nuanced and more complete proposition linking coalition power to major policy change. The reformulated hypothesis states, "A shift in the balance of coalition hard power, a shift in the balance of coalition soft power, imposition by a hierarchically superior jurisdiction, or some combination thereof, are necessary, but not sufficient, sources of change in the policy core attributes of a governmental program."

This chapter investigates this hypothesis in the Alberta water policy subsystem by examining the hard and soft power resources in the subsystem and whether there were shifts in the balance of either type of power. The expectation, also outlined in chapter 2, is that there was little change in the balance of hard power, but that there was a substantial shift in the balance of soft power towards the Green coalition, providing it with the influence it needed to persuade Aggies and government decision-makers of the need to adopt soft water policy reforms. Since the Canadian government refused to involve itself in Alberta's domestic water policy, imposition by a higher-level government is ruled out as a cause of Alberta's water policy reform. In this way, it is hypothesized that it was a shift in soft power that most contributed to the adoption of water reforms.

The hard and soft power resources of the Aggie and Green coalitions in the water policy subsystem are investigated below. The first section examines the coalitions' hard power resources (the formal legal authority to make policy decisions), while the second section surveys the coalitions' soft power resources (public opinion, information, mobilizable troops, and skilful leadership). Financial resources are included as a soft power resource, since they have been most often used in this manner. Each resource is investigated for the entire period of study, from the early 1970s to the mid-2000s, and is evaluated using a combination of descriptive statistics and qualitative evidence.

Hard Power Resources

Formal-Legal Authority to Make Policy Decisions

It is no secret that irrigators in southern Alberta have long been a group with considerable electoral and legislative clout. The sources of this power can be traced to four constitutional and political factors that, in fortuitous combination, have granted the Aggies considerable hard power in the water policy subsystem. First, the Canadian constitution and federal policy, at least since the early 1970s, have given the Alberta government almost exclusive jurisdiction over water and water development in southern Alberta. Second, irrigators and irrigation-dependent communities are geographically concentrated in the southern part of the province, and that position, in the prevailing single-member plurality electoral system, has given the Aggies considerable electoral influence over the swathe of seats in the irrigation belt. Third, as the result of Alberta's seat distribution practices, rural areas have been considerably over-represented in the legislature compared to urban areas, amplifying the electoral and legislative importance of the irrigation belt. Finally, from 1975 to 2012, the irrigation belt consistently returned PC MLAs to the Alberta legislature, giving them strong representation in the party that governed the province continuously from 1971 to 2015. Together, these four factors have enabled the Aggies to place members within Alberta's governing caucus and Cabinet, giving them authoritative policy influence that the Greens have simply been unable to match.

As described in chapter 3, both the federal and provincial governments were involved in the allocation and development of southern Alberta's water for the first three-quarters of the twentieth century, but federal involvement gradually dwindled until it retreated altogether in the early 1970s. The key decision, in this regard, was Ottawa's move to extricate itself from irrigation development in 1973, transferring most of its irrigation responsibilities and assets to the provincial government (Irrigation Water Management Study Committee 2002, 17). From this point forward, Canadian governments showed almost no inclination to get involved in intra-provincial water issues.

Much of this had to do with the decentralized nature of Canadian federalism and the determination with which the Canadian provinces protected their jurisdiction over natural resources. Alberta was a particularly fierce defender of this jurisdiction – particularly after the ill-fated National Energy Program of the early 1980s – and any federal meddling with Alberta's resources risked alienating both the Alberta government and the Alberta electorate. Consequently, in water issues, the federal government confined itself to dealing with matters that had an explicit international or inter-provincial dimension or matters that clearly touched on its jurisdiction related to fisheries, Indian lands, navigation and shipping, or environmental assessment (Johns and Rasmussen 2008). But even in these instances, Ottawa stepped very lightly and generally took the approach that its role in water allocation stopped at the Alberta border. This, in effect, created a single-venue water policy subsystem in southern Alberta, with almost all water policymaking authority at the provincial level (Irrigation Water Management Study Committee 2002, 17; de Loë 1994, 201).

This provincially centred, single-venue policy subsystem was much to the advantage of the Aggies and much to the disadvantage of the Greens. The Aggies were advantaged because they were politically much stronger at the provincial level than at the federal level. At the provincial level, the Aggies held sway over the seats in the irrigation belt and had forged a close political relationship with the governing PCs, giving them substantial political influence in the Alberta government

(Glenn 1999, 130–40). At the national level, the electoral and legislative influence of Alberta's irrigators was considerably diluted and they had not forged a close political relationship with any of the national parties, so the confinement of water allocation issues to the provincial level played to their strengths. The Greens, in contrast, had nowhere near the electoral or legislative influence of the Aggies at the provincial level, and the absence of a federal water policy venue effectively pre-empted an important strategy used by many minority coalitions to influence policy development: venue shifting. The Greens could not make end runs around the Aggies' provincial political power by taking water issues to Ottawa because Ottawa would not touch them (Heinmiller 2014). The Greens tried to do this when Friends of the Oldman River took court action to force the federal government to apply its environmental assessment process to the Oldman Dam, but the Alberta government ignored the federal panel's report and Ottawa was completely unwilling to enforce the panel's recommendations. Thus, the Aggies and Greens were confined to the provincial policy venue, where the Aggies had considerable electoral and legislative clout.

One fundamental reason for the Aggies' electoral power is their geographic concentration. Just about all of Alberta's irrigation is concentrated in the SSRB, and all of Alberta's irrigation districts are concentrated in the area south and east of Calgary. In the context of a single-member plurality electoral system, this has meant that most of Alberta's irrigators, irrigation-related businesses, and irrigation-dependent communities are concentrated in a relatively small number of electoral constituencies known as the irrigation belt. Because the Aggie vote is concentrated in the irrigation belt, the Aggies have had considerable influence over these seats and have used this influence to elect MLAs who will advocate the Aggie cause. The Greens, in contrast, are geographically dispersed throughout the province and, even though they are sizeable in number, have been unable to achieve the critical mass of votes needed to gain control over specific ridings. The tendency of single-member plurality electoral systems to over-reward geographically concentrated interests and under-reward geographically dispersed interests is well-established in the political science literature, and the Aggie and Green advocacy coalitions in Alberta are classic examples of this phenomenon (Cairns 1968).

With effective electoral control over the seats of the irrigation belt, the Aggies have also benefited from Alberta's seat distribution practices, which have consistently over-represented rural areas in the

Alberta legislature, giving the irrigation belt more seats than it would deserve on the basis of its population only (Brownsey 2008, 150). There is a considerable literature documenting the long history of rural over-representation and urban under-representation in the Alberta legislature (Long 1969; Pasis 1973; Archer 1993; Neitsch 2011). Archer (1993) has shown that rural areas have always been over-represented in Alberta, but the disparity increased in the 1950s and 1960s as the assignment of urban seats did not keep pace with urban population growth. Neitsch (2011) further argues that rural over-representation was preserved between the early 1970s and mid-1990s, as the PCs, who enjoyed strong rural support, worked to preserve rural over-representation each time the seat distribution rules came up for reform.

The extent of rural representation and the electoral clout of the irrigation belt are illustrated in table 5.1 below. The first thing to note is the sizeable chunk of seats in the Alberta legislature constituting the irrigation belt. Even by 2008, the irrigation belt still comprised 14 per cent of the seats in the legislature, a substantial number for any party leader trying to put together an electoral coalition for a majority government. Another thing to note is the relatively small decline in irrigation belt seats between 1975 and 2008, a period in which Alberta's cities and northern hinterland were growing by leaps and bounds. This hints at the rural over-representation in the Alberta legislature and suggests that the irrigation belt, in particular, has been a beneficiary of this practice. However, it is also important to point out that, over the long term, the legislative clout of the irrigation belt has been in decline and this trend seems likely to continue. Alberta's cities and suburbs are still growing fast, creating demographic and political pressures to allocate more seats to urban areas and to correct urban under-representation. So while the Aggies will continue to enjoy the benefits of a geographically concentrated interest in a single-member plurality electoral system, their overall legislative influence will likely continue to wane.

While the irrigation belt seats gave the Aggies considerable legislative influence, the irrigation belt electors' consistent support for the governing PC Party gave the Aggies considerable executive influence, as well. Albertans are said to prefer "non-partisan" politics, featuring dynastic political parties elected to office for long periods of time, and a style of politics in which the most significant political cleavages are often played out within the dominant party rather than across party lines (Macpherson 1953; Pal 1992). Since its admission to Confederation, Alberta has had four successive governing dynasties, starting with the Liberals

Table 5.1 Legislative Seats in the Irrigation Belt, 1935–2008

Year	Irrigation belt seats*	Total seats	%
1935	14	63	22.22
1955	12	61	19.67
1975	13	75	17.33
1993	12	83	14.46
2008	12	83	14.46

Source: Heritage Community Foundation (n.d.).
* The irrigation belt is defined here as all ridings south and east of Calgary, excluding greater Calgary but including the cities of Medicine Hat and Lethbridge.

(1905–21), then the United Farmers of Alberta (1921–35), Social Credit (1935–71), and the PCs (1971–2015). In order to have policy influence, it is important to have inside access to the current governing dynasty. MLAs in the opposition ranks are particularly ineffectual in Alberta politics and rarely have significant influence on public policy outcomes.

Voters in the irrigation belt have been particularly canny about this dynamic of Alberta politics and have supported every governing dynasty in the province's history. With respect to the PCs, the irrigators were a bit late in this support, having voted overwhelmingly for Social Credit in the 1971 election that saw the PCs first come to power. However, in the 1975 election the irrigation belt switched en masse, with eleven of thirteen irrigation belt seats going to the PCs. Over the following eight elections, more than 90 per cent of the seats in the irrigation belt went to the PCs, a trend that was broken only in the 2012 election when the irrigation belt abandoned the PCs in favour of the Wildrose Party (Heritage Community Foundation n.d.; *Canadian Election Atlas* n.d.). The irrigators continued to support Wildrose in the 2015 election, bucking the "orange wave" that resulted in the province's first NDP government.

For their part, the PCs did their best to cultivate a strong relationship with the irrigators, particularly as the party lost support in some urban areas, such as Edmonton, and relied on rural support as an important part of their electoral coalition (Neitsch 2011). With the irrigation belt consistently well-represented in the governing caucus, a number of these MLAs were appointed to Cabinet positions, in portfolios such as Finance, Agriculture, and Environment. This gave the Aggies important access to executive decision-making processes and at least some formal legal authority over policy decisions.

The Greens have also attempted to make legislative inroads in Alberta politics, but with very little success. The Green Party of Alberta was formed in 1986 and contested every election up to 2008. In that time, it never elected an MLA, never received more than 4.55 per cent of the popular vote, and fielded a full slate of candidates only once (*Canadian Election Atlas* n.d.). By 2009, the party was in such disarray that it failed to submit its annual financial statement to Elections Alberta and was deregistered. The party has since been resurrected, but is hardly a political force (Green Party of Alberta 2012). To be fair, the party has been severely handicapped by the single-member plurality electoral system, which penalizes parties with low levels of geographically dispersed support. The bottom line, though, is that there has been no mechanism available to convert the environmentalist vote into any kind of legislative or executive power, so the Green advocacy coalition has never had any kind of formal legal authority to make policy decisions.

Overall, in terms of electoral clout, legislative influence, and executive presence, the Aggies have far outstripped the Greens and have managed to gain some formal legal authority over water policy while the Greens have not. This formal legal authority has come from having Aggie coalition members in the Alberta legislature, the governing party caucus, and in Cabinet. The only high office where the Aggies have not placed one of its members is the premiership. This absence is notable, particularly given the large literature documenting the unique power and unrivalled policy influence that first ministers have within Canadian governments (Savoie 1999, 2010; Bernier, Brownsey, and Howlett 2005). For this reason, it is worth briefly exploring the ideological proclivities of the three Alberta premiers who held office during the period of study, Peter Lougheed, Don Getty, and Ralph Klein, to determine whether they may have had Aggie or Green predilections.

Peter Lougheed was a titan of Alberta politics, the person most responsible for ending the Social Credit dynasty and starting the PC one. He gained four successive majority governments, serving from 1971 to 1985, and became an important national figure during the natural resource conflicts and mega-constitutional politics of the late 1970s and early 1980s. Ideologically, Lougheed has been described as a "strong-state Tory" whose overriding priorities were province-building and economic diversification (Tupper 2004). These priorities aligned quite well with the Aggies' irrigation expansion agenda, which also relied on extensive state involvement in support of private enterprise. Lougheed had few rural connections himself, but his province-building approach was effectively linked to irrigation development by his ministers of agriculture, particularly Hugh Horner (1971–5) and LeRoy Fjordbotten (1982–6), both of whom were from southern Alberta. Lougheed also presided over the early development and expansion of the province's Department of the Environment, which he had inherited from his Social Credit predecessors. Particularly later in life, he was also known for his environmental consciousness and spoke out on a number of environmental issues, including the rapid exploitation of the northern oil sands. Overall, it is not possible to place Lougheed clearly in either the Aggie or Green camp, and it seems likely that he would have been open to arguments from both coalitions.

Don Getty was premier of Alberta from 1985 to 1992, a particularly turbulent time for Alberta agriculture. Getty continued Lougheed's interventionist approach to governance, an approach that seemed to become less effective and less popular throughout his term in office. Even though he did not have rural roots himself, Getty had strong sympathies for the agricultural community: rural areas were a key part of his electoral coalition, and, after losing his Edmonton seat in the 1989 election, he went on to represent the rural riding of Stettler in central Alberta through a by-election. Getty presided during a time of depressed commodity prices and a period of drought in the mid-1980s; in response, his government spent well over $1 billion in propping up Alberta agriculture. Getty did not appear to have any particular environmentalist sympathies, but, as a government manager, he was also inclined to let ministers manage their own departments. In 1989, he appointed Ralph Klein as environment minister, a particularly conflict-ridden portfolio at the time, and he tolerated and supported Klein's efforts to introduce a number of environmental reforms (Lisac 2004). On balance, Getty was probably more sympathetic to the Aggies than the Greens, but he was certainly not a member of the Aggie coalition.

Ralph Klein was another titan of Alberta politics, serving as premier from 1992 to 2006 during four successive majority governments. Klein had a multifaceted ideology that is difficult to define: he was mayor of Calgary during most of the 1980s but was enormously popular in rural areas; he was a strong environment minister but a tireless business proponent; and he was a suspected closet-Liberal but steered his first government in a radically neoconservative direction. Klein was the first former environment minister to become premier and, as environment minister, "Klein brought to Cabinet a message that a new public concern about environmental matters had to be accommodated"

(Lisac 2004, 248). However, it was overwhelming support from rural Albertans that allowed him to win the race for the PC leadership in 1992, and rural seats were an important part of his electoral coalition (Barrie 2004, 261). Klein's governments also took a hard ideological turn to the right, and neo-conservativism was an important influence, particularly during the first Klein government. Ministers such as Murray Smith, Lorne Taylor, Ed Stelmach, Jon Havelock, and Lyle Oberg were particularly forceful in pushing the neoconservative agenda (Martin 2002, 142). Like Lougheed and Getty, it is impossible to place Klein in either the Aggie or Green coalitions, though he was probably more amenable to the arguments of the Greens than any of his predecessors.

From these brief synopses, it is clear that none of the three premiers during the period of study can be squarely placed in either of the prevailing advocacy coalitions. This means that the premiers, the most important of all policy decision-makers, were open to influence in their water policy decisions, with both advocacy coalitions vying to sway them. This makes the presence of Aggies in the legislature, the governing caucus, and the Cabinet all the more important. These policy elites were well-placed to make policy decisions, in some instances, and to influence other policy decision-makers, in other instances. This inside access to the centre of policymaking is something that the Greens entirely lacked, and marks the Aggies as having a decisive advantage in hard power, even if that advantage has dwindled somewhat over time.

Soft Power Resources

Financial Resources

As described above, financial resources are one of the most important resources that an advocacy coalition can have. A full consideration of a coalition's financial resources must take into account two factors. First, one must consider the importance of the coalition members and their economic activities in regional, provincial, and national economies – their economic structural power. This is necessary because coalitions engaged in activities important to regional, provincial, or national economies can mount very persuasive arguments to policy decision-makers in terms of stimulating economic growth, providing employment, and increasing general prosperity. The second factor is the capacity of the coalition to capture or raise revenues that might be used for policy advocacy. This is obviously important because it indirectly determines what advocacy campaigns a coalition can fund and what advocacy services, such as professional lobbyists, they can purchase. During the period of study, the Aggies enjoyed a considerable advantage over the Greens on both factors.

While Alberta was once primarily an agricultural economy, the relative importance of agriculture in the provincial economy has continuously shrunk since the 1930s. At that time, agriculture accounted for almost 60 per cent of Alberta's GDP (Government of Canada 1930). Then, in 1936, oil was discovered in the Turner Valley, an oilfield that, within five years, turned out to be the largest in the British Empire. Even more oil was discovered near Leduc in 1947 and, with this discovery, Alberta's oil sector really took off. Though the expansion of the oil sector was far from linear – proceeding through a number of booms and busts – the overall transition of Alberta's economy from an agricultural base to an energy base was unmistakable. Agriculture did not go into decline – in fact, it continued to grow – but it was simply outpaced by the large and wealthy oil sector, so much so that, by the mid-1980s, agriculture's share of provincial GDP had fallen to only 3 per cent and has continued to shrink since (Statistics Canada n.d.; Government of Alberta, "The Economy" 2014).

For irrigators, this economic structural change has meant that their industry has gone from being one of provincial importance to one of regional importance. It is estimated that 13 per cent of (southern) regional gross domestic product, 19 per cent of regional production, and 30 per cent of regional employment opportunities are directly or indirectly associated with irrigated agriculture. This translates into about $5 billion worth of production per year and about 13,000 full-time jobs. It is also estimated that, without irrigation, the regional population could be reduced by as much as 75 per cent, as land is taken out of production, farms are consolidated, and irrigation spin-off jobs disappear (Alberta Agriculture, Food and Rural Development 2004). In short, irrigation is relatively insignificant on a provincial scale, but, on a regional scale, it is very important in terms of both jobs and production.

The regional economic importance of irrigation has provided an important soft power resource to the Aggies in Alberta's water policy debates. Any attempts to curtail irrigators' access to water or to reduce public expenditures on irrigation support have been met with stark warnings from the Aggies about the negative impacts of such measures on jobs, production, and prosperity. Moreover, these effects would be concentrated in one region, a region still haunted by the memories of massive rural depopulation during the droughts and commodity price crashes of the 1920s and 1930s (Jones 1987). Aggies have also argued that continued irrigation support is important for economic diversification, particularly as water is a renewable resource but oil is not. All of these arguments carry considerable weight with policy decision-makers, but they are also tinged with self-interest in a way that the Greens' arguments about environmental protection and restoration are not. The self-interest intertwined with the Aggies' economic arguments can sometimes diminish their persuasiveness, but these economic arguments are still one of the primary means the Aggies use to influence policy decision-makers.

Another substantial change in the Alberta economy has been the growth in importance of tourism. By the mid-1980s, tourism and consumer services constituted a larger share of provincial GDP than even agriculture (Government of Alberta, "The Economy" 2014). The tourism industry is spread throughout Alberta, but some of the most attractive tourist destinations are in the south, are based in the natural environment, and depend largely on river water. The Banff and Kananaskis resort areas on the upper reaches of the Bow River, Waterton Lakes National Park in the headwaters of the St Mary River, and the recreational fisheries along the Bow and South Saskatchewan Rivers all depend on adequate water flows to sustain them, and their contributions to local and provincial economies are substantial. This has provided the Greens with an economic argument for preserving and restoring natural flows on these rivers, helping to offset the economic arguments of the Aggies, at least to some extent.

When converting their economical structural power into money that can be used to fund advocacy activities, the Aggies have had a decisive advantage over the Greens. One way to gauge this is to look at the budgets and fundraising capacities of the umbrella organizations in each coalition. In the Aggie coalition, the Alberta Irrigation Projects Association assesses annual membership levies from the irrigation districts on a per acre basis; in 1996, for example, this levy was eighteen cents per acre. This gave the association an annual budget of around $245,000, 40 per cent of which went to administrative salaries and overhead for its permanent office in Lethbridge. The other 60 per cent – approximately $148,000 – went directly to public relations, education, and lobbying (Klassen 1996, 4–5). These figures do not include money spent by other Aggie umbrella organizations such as Unifarm, the Alberta Cattle Commission, and the Southern Alberta

Water Management Committee, who also spent money on various water policy advocacy activities.

In contrast, the Greens do not have industry actors from whom they can regularly collect levies. Most Green groups are charitable organizations that rely on donations and, according to federal legislation, charitable organizations can spend only 10 per cent of their charitable tax receipted money on political advocacy. Since tax receipting is one of their most effective ways of raising money, this seriously constrains Green groups' funding of political advocacy (Stefanick 1991, 75). Reliance on donations also means that the Greens have to get by on much less than their Aggie counterparts. For example, by mid-1993, Friends of the Oldman River had raised about $200,000 in donations during its entire existence. This does not count all of the pro bono services provided to it by lawyers, but is still less than the annual operating budget of the Alberta Irrigation Projects Association (Glenn 1999, 148). In the early 1990s, the Alberta Environment Network had annual expenditures of around $110,000, but all of this was spent on coordination and communication activities, as the network does not fund or get involved directly in policy advocacy (Stefanick 1991, 94).

Overall, the Aggies enjoyed a privileged position when it came to financial resources during the period of study. The regional economic importance of irrigation gave them substantial economic structural power, only somewhat offset by the growth of tourism and its reliance on the natural environment. Moreover, the Aggies had much more capacity than the Greens to raise funds to support policy advocacy activities. Though the Aggies' advantage in financial resources is clear, it is important to note that this was the only soft power resource in which they enjoyed an unmitigated advantage.

Public Opinion

Given Alberta's reputation – or stereotype – as a bastion of conservatism and frontier free enterprise, it may be surprising to learn that public opinion was one of the most important resources available to the Green coalition. The reasons for this are complex and manifold. Part of it had to do with the public education and advocacy efforts of the Greens themselves who, lacking influence inside government, worked hard to cultivate public opinion on water issues and place pressure on government decision-makers from the outside. The more deep-seated reasons for the Greens' favourable public opinion had to do with long-term social and demographic changes within Albertan society, such as immigration, urbanization, and the rise of post-materialist values.

Most Western societies experienced sharp increases in population, immigration, and urbanization during the second half of the twentieth century, but these changes were especially rapid and pronounced in Alberta, as they were stimulated by the province's prolonged oil boom. Between 1951 and 2011, the population of Alberta increased from 940,000 to 3.6 million people. Most of this population increase came from immigrants, from other countries and other Canadian provinces, drawn to Alberta in search of jobs in the energy sector. During the same period, the percentage of the population living in urban areas increased from 48 per cent to 79 per cent (Arcand and Lefebvre 2012). Alberta urbanized later than most other jurisdictions in the developed world, but the extent of its urbanization has now gone farther than most other Canadian provinces (Archer 1993). Albertans remain proud of their rural heritage, as evidenced in the continuing popularity of the Calgary Stampede and the presence of cowboy hats and boots on the streets of Calgary, but Alberta is overwhelmingly an urban society.

The effects of these demographic changes on the water policy subsystem have been profound. As Alberta urbanized, an increasing number of voters came to live in urban and suburban areas, and the relative political importance of cities has been on the rise. Urbanization, immigration, and the economic importance of the oil industry have all combined to ensure that fewer and fewer Albertans are directly connected with and reliant on agriculture, resulting in less overall public interest and sympathy with rural concerns. Albertans in general have also become much wealthier over the last half-century, and research has posited a connection between Western societies' increasing wealth and the rise of post-materialist values such as environmentalism (Inglehart 1990). In short, by the late 1980s, Albertans had become much more urban, much less connected to agriculture, and much wealthier than in the past, all of these forces converging to make them much more receptive to the Greens' message.

As Paehlke has shown, environmentalism in Canada has experienced at least two waves of public support, the first in the late 1960s and early 1970s, and the second in the late 1980s and early 1990s. During these waves, public support for environmentalism and environmental protection has peaked, only to subside to a baseline level until the next wave (Paehlke 2000). The waves usually develop during good economic times when the public can afford to focus on post-material issues such as the environment, and crash in the midst of an economic downturn when public attention returns to more "bread and butter" issues. Each wave has left important legacies with lasting effects. The legacy of the first wave in Alberta was the creation of many of the original conservationist and environmentalist groups as well as the formation of the Department of the Environment. The second wave also left a number of important legacies, including the coalescence of the Greens into a bona fide advocacy coalition, as discussed in chapter 4. It also coincided with the Oldman Dam controversy, focusing public attention on water issues and generating substantial public concern with protecting the province's water resources, a concern that was sustained beyond the end of the second wave.

Polling conducted by the Alberta government during and after the second wave illustrates the persistent salience of water resource protection for Albertans.[7] In 1992, the Department of the Environment, the Environment Council of Alberta, and the Water Resources Commission conducted a survey involving almost six thousand randomly distributed questionnaires. The original survey data have been lost, but a summary document describes the results:

- People are very concerned about water management in Alberta and believe it is a very important issue.
- People are relatively unfamiliar with water management legislation.
- People want government decision-making, control, and public participation to become more effective. People do not support a "wait and see" attitude. Some indicate that they want the issues addressed effectively and immediately.
- People are in favour of water metering.
- People will consider putting a price on water for the quantity used and degradation of water quality, as an avenue for further exploration.
- People expect they may have to pay to preserve the quantity and quality of water.
- People say they are prepared to become personally involved in the policy and legislation review process. (Alberta Environment, Environmental Council of Alberta, and Water Resources Commission 1992, 104–5)

In the early 2000s, another survey by Alberta Environment, the results of which are summarized in table 5.2, showed that water issues were

7 Environics polling during the same period supports this claim, though the sample sizes from Alberta were too small to be statistically significant.

Table 5.2 Polling Data on Water Issues, 2002

Statements	Concerned (%)	Very concerned (%)
Industrial and agricultural growth will increase demand for water in places where water is not always plentiful.	30	38
Water will become an increasingly scarce resource in Alberta.	21	40
Climate change will seriously affect water flows in Alberta.	22	30
Albertans will have to choose between sustaining aquatic ecosystems or economic growth.	22	23
Effective water management will require a lot more investment in infrastructure, such as for dams, and for the maintenance of that infrastructure.	24	23
Statements	**Agree (%)**	**Strongly agree (%)**
The province should prepare for the possibility of water scarcity in the province.	30	61
If there are water shortages in the future, the province should put a higher priority on preserving natural aquatic environments, even if this limits economic growth and jobs.	38	50
The province should explore ways to increase water conservation, even if this increases costs to the person using the water.	40	42

Source: Alberta Environment ("Water for Life – Results" 2002).

still salient and that there was still strong public support for water protection measures. One of the most notable findings was that 88 per cent of Albertans agreed or strongly agreed that preserving aquatic environments should be given priority in instances of water scarcity, even if it means limiting jobs and economic growth. This sort of environmental consciousness is at odds with the stereotype of typical Albertans and reveals that the Greens enjoyed considerable public support for their policy positions. To the extent that Alberta's water policy decision-makers were aware of this public support, the Greens were able to use it as a soft power resource to pressure policymakers into adopting at least some of their water policy preferences.

Information

One of the most important resources the Greens have possessed is information, particularly scientific information. Scientific studies supporting an advocacy coalition's claims can provide credibility to a coalition's policy positions, can be used to discredit opposing coalitions, and can persuade policy decision-makers of the inherent "rightness" of a particular policy. Moreover, on environmental issues, polling has shown that the public tends to trust the claims of scientists more than others. A 1994 poll showed that over 84 per cent of Canadians believed some or most of what scientists and academics said on environmental issues, while only 30 per cent believed some or most of what industries and businesses said (Ipsos-Reid 1994).

In the water policy subsystem, the information that mattered most was that on the capacity and health of the heavily used rivers of the SSRB. Throughout the period analysed here, many scientific studies of the SSRB's water resources were undertaken, both by government agencies and by academics. The results of these studies painted an increasingly dire picture of the SSRB's environmental health, and many studies recommended specific policy measures – most of them soft – to halt the rivers' decline. As these findings and recommendations were much closer to the Greens' policy positions than the Aggies', they provided the Greens with considerable rhetorical ammunition in their exchanges with Aggies and policymakers.

Until the 1980s, most studies of the SSRB were undertaken with dam construction and water supply management in mind. For instance, the massive Saskatchewan-Nelson Basin Study undertaken by the Prairie Provinces Water Board between 1967 and 1976 identified nineteen potential dams and eleven potential diversions in Alberta, and a series of studies by the Alberta government between 1975 and 1978 investigated a variety of potential dam sites on the upper reaches of the Oldman River. These studies paid scant attention to the health of southern Alberta's rivers and were focused squarely on developing more water supplies to meet growing water demands. Then, in the early 1980s, the focus of government studies began to shift: rather than evaluating suitable dam sites, more attention was given to evaluating instream flows needs. As a result of these studies, the efficacy of the hard policy approach began to be questioned. This shift was subtle at first, but became more pronounced over time. By the early 1990s, government reports were outright recommending the abandonment of the hard policy approach and, by the early 2000s, were advocating the closure of the SSRB to new water licences. These recommendations unequivocally supported Green policy positions and were a great boost to their advocacy efforts.

The first studies to focus squarely on instream flow needs in the SSRB were undertaken by the Fisheries Management Division of Alberta Environment, starting in 1982. These studies were undertaken in support of water use activities such as the protection of local sport fisheries, the renewal or approval of infrastructure licences, and as part of the department's water planning process for the SSRB. Between 1982 and 1998, instream flow needs studies were completed for all or parts of the Bow River (1990–4), Elbow River (1994–5), Highwood River (1984–90), Kananaskis River (1996–7), Pekisko Creek (1986), Sheep River (1982–3), Smith-Dorrien Creek (1991–2), Belly River (1987–94), Oldman River (1988–90), St Mary River (1987–94), Waterton River (1987–94), Willow Creek (1991–2), and Red Deer River (1991–8) (Instream Flow Needs Committee 1998, D1–D13). Though these studies generally had utilitarian purposes, they also focused attention on the health of riverine environments, provided more data on the state of these environments, and offered some rough indications of what would be needed to protect or restore these environments, all of which fed into future water policy processes.

One of the most important studies in the history of Alberta water politics was that by the Oldman River Dam Environmental Assessment Panel in 1992. Friends of the Oldman River fought a protracted court battle, all the way to the Supreme Court of Canada, to force the reluctant federal government to apply its environmental assessment process to the Oldman River Dam project. When the panel finally reported in 1992, it not only rejected the dam and recommended that it be decommissioned, it also repudiated the entire hard approach to water governance: "Should all the proposed acres allocated for irrigation be developed, the panel fears that the situation [of water shortage] would quickly become the same as it was in the early 1980s and pressures would develop for more dams and diversions to meet this 'need' for more water. As long as water is provided to users without charge, and environmental protection is undervalued, as it was in the planning for this project, more environmentally damaging projects will be proposed. The panel very much wishes to avoid such a future" (Oldman River Dam Environmental Assessment Panel 1992, 12).

Though the report did not stop the construction of the Oldman Dam, it dealt a heavy blow to the credibility of the hard policy approach and provided a credible informational resource to the Greens.

A decade later, another landmark report changed the terms of the water policy debate even further. This was the "Instream Flow Needs

Determinations for the South Saskatchewan River Basin, Alberta, Canada" report, prepared between 2001 and 2003. The report was compiled by a team of researchers from various government departments, in consultation with private sector experts, and was included as part of the background documents for the development of the second phase of the SSRB Water Management Plan. These background documents were distributed to all of the stakeholders involved on the basin advisory committees, the bodies negotiating and recommending plans for each of the SSRB's four sub-basins, and played a key role in framing their discussions. Using instream flow needs indicators based on fisheries habitat, water quality, riparian vegetation, and channel maintenance, the report clearly showed that the three southern sub-basins were already allocated beyond their instream flow needs (Clipperton et al. 2003, iii). The evidence was so compelling that it was accepted by both Greens and Aggies, bringing the Aggies much closer to the Greens' conception of the policy problem, and laying the groundwork for a consensus on closing the Bow, Oldman, and South Saskatchewan sub-basins to new water licences.

Dire reports on the state of the SSRB's rivers were coming not only from government experts, but from academia as well. From the 1970s onward, the environmental sciences became popularized as academic disciplines and environmental research expanded greatly in both depth and breadth. In Alberta, considerable research was focused on the SSRB and just about all of this research told a story of environmental decline. Two researchers, in particular, were emblematic of this: Stewart Rood at the University of Lethbridge and David Schindler at the University of Alberta. Rood specialized in the study of Prairie riparian forests, mostly poplars and cottonwoods, and documented their decline due to dam construction and other interruptions in natural streamflows and flooding events (Rood and Heinze-Milne 1989; Rood and Mahoney 1990; Rood et al. 1995). Schindler specialized in water quality issues, conducting extensive experimental research on lake eutrophication from agricultural fertilizer discharges, a growing problem in downstream Lake Winnipeg, as well as pesticide and PCB pollution in western Canada's glaciers and rivers (Schindler and Lean 1974; Blais et al. 2001).

By the early 2000s, both Rood (Rood et al. 2005) and Schindler (Schindler and Donahue 2006) were warning about the over-allocation and environmental decline of the SSRB, as well as the negative implications of climate change. In their research, Schindler and Donahue showed an 83.6 per cent reduction in summer flows between 1912 and 2003 in the

Figure 5.1 Historical Summer River Flows in the SSRB, at Saskatoon

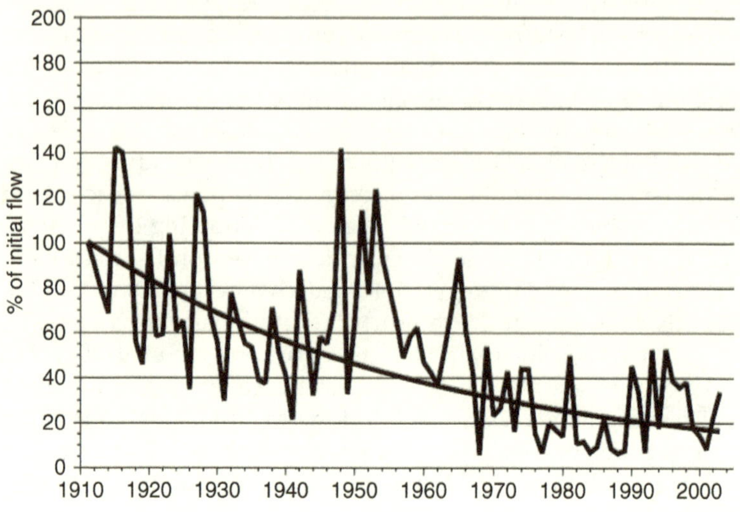

Source: Schindler and Donohue (2006, 7212).

South Saskatchewan River, much of it due to irrigation water extractions (see figure 5.1). They also warned of further human-induced reductions in the available water supply due to the effects of climate change. They presented data showing that the glaciers in the headwaters of the SSRB had shrunk by about 25 per cent over the past century and that much more winter precipitation is likely to fall as rain, reducing the snowpack and shrinking river flows during the spring freshet (Schindler and Donahue 2006, 7212). They also put the twentieth-century climate into longer historical context through an analysis of past climate proxies such as tree rings and salinity-sensitive diatoms. They found that the twentieth century had been unusually stable and moist and that "even the droughts of the 1930s and the past few years [1998 to 2004] are relatively mild when considered in historical context" (7210).

These academic studies, when added to the government studies already recounted, reinforced the Greens' argument that the SSRB was in serious decline and that substantial policy action was needed to arrest this decline, to reverse it, and to prepare for the future challenges of climate change. For a coalition lacking hard power and relying on shoestring financial resources, this sort of informational credibility was crucial. It was one of the factors that allowed them to cultivate consistently favourable public opinion, and it had important persuasive effects on policy decision-makers and, at times, even the Aggies.

Mobilizable Troops

The one resource in which neither advocacy coalition seems to have had a clear advantage was in mobilizable troops. Mobilizable troops are the people that a coalition can rally to take part in campaigns to support their cause, whether in election campaigns, public education campaigns, or public protests. The Greens relied on the mobilization of troops more than the Aggies, mostly because the Greens lacked inside access to policy decision-makers and tried to use the mobilization of troops to pressure policy decision-makers from the outside. The Greens also relied on the mobilization of troops substantially at first, but less so, over time. The origin of the Greens' collective action was in their resistance to the Oldman River Dam, and public protests and public education campaigns were an important part of this resistance. Once the Greens became recognized stakeholders in the water policy subsystem and were regularly invited to participate in policy consultation processes, they resorted to public demonstrations much more sparingly.

Though it is difficult to determine the precise number of mobilizable troops available to each coalition, best estimates suggest that their numbers were relatively comparable. Estimates were calculated by looking at the organizational membership numbers for all known coalition participants (as determined in chapter 4) as well as those organizations that were closely related to each coalition but were only occasional policy process participants. The intent was to try to get an idea of how many people each coalition could mobilize in defence of its most cherished causes. Although the resulting estimates are rough, they are useful in that they do not show a clear advantage to either coalition.

The Aggies' mobilizable troops easily numbered in the tens of thousands, including the irrigators themselves and those dependent in one way or another on irrigation. There are roughly nine thousand irrigators in southern Alberta, around six thousand organized collectively in the irrigation districts, and around three thousand holding private water licences outside of the districts. There are also a number of food processing plants in the region that rely on irrigation-produced commodities, including companies such as Canbra Foods, Empress Foods, Lamb Weston, McCain Foods, Rogers Sugar, and Spitz. The employees of these companies, along with those from agricultural equipment and supply dealers in the region, number in the thousands. In addition, Unifarm and many of its four thousand members could also have been rallied to the Aggie cause. Then there are the local communities in which the irrigators live and work. These numbers are the most difficult of all to estimate, but it is indicative that over fifty communities in southern Alberta have their domestic water supplies delivered through the dams and canals of the irrigation system, and that many communities in the region either have a strong stake in irrigation or are entirely irrigation-dependent. This creates a large pool of potentially mobilizable troops, adding to the Aggies' numbers even further.

The Greens' mobilizable troops also numbered in the tens of thousands, with most of them coming from environmentalist and conservationist groups concerned with water governance. Focusing on the year 1995 – a midpoint in the period of analysis and one of the years examined in depth in chapter 4 – an estimate of the Greens' mobilizable troops can be calculated using the group membership figures provided in the *Alberta Environmental Directory*. By summing the membership numbers of all known 1995 Green coalition groups (identified in chapter 4) and all groups who self-identified in the 1995 directory as having an interest in water quantity or quality in the southern Alberta region, the total number of Green mobilizable troops comes out to just over 45,000 people (Pembina Institute for Appropriate Development 1995). This is a very impressive number, but is probably an overestimate of Green mobilizable troops for a couple of reasons: first, some individuals in this estimate are double- or triple-counted, given that they were members of more than one Green group; and second, a substantial number of individuals in this estimate would not have been mobilizable either because they lived in central or northern Alberta and had only a peripheral interest in southern water issues, or they joined Green organizations to work on local conservation projects and were largely non-political. Accordingly, the 45,000 figure should probably be regarded as a maximum number of mobilizable troops, with the more likely number being somewhat lower. An examination of the 1989 and 2002 editions of the directory also shows that Green membership numbers trended moderately upward over time, suggesting a modest increase in the number of mobilizable Green troops throughout the period of study.

While the estimates of Aggie and Green mobilizable troops presented here are somewhat rough, it is safe to conclude that the number of troops available to each coalition was in the same order of magnitude and that neither coalition had a clear advantage in this resource. Though the Greens relied on their mobilizable troops more than the Aggies, the Aggies' mobilizable troops were also important because they were, in effect, a resource in reserve, mobilized only when needed, such as when the Aggies resisted the initial draft of the Water Act in 1994–5 (as we will see in chapter 7). In short, mobilizable troops were a resource available to both coalitions, the coalitions were comparable in mobilizable troops, but the Greens relied on them more, mostly as a means of compensating for their deficiency in other resources.

Skilful Leadership

The least tangible and measurable of all coalition resources is skilful leadership, but its importance should not be underestimated. Well-connected, energetic, and charismatic leaders can help to augment a coalition's other resources, by mobilizing troops, cultivating public opinion, lobbying policy decision-makers, and expanding fundraising. They may also be the glue helping to keep a coalition together, mediating and resolving disputes between intra-coalition factions. Some coalition leaders may also be master political tacticians, deftly out-manoeuvring competing coalitions in policy development, sometimes with fewer resources than their opponents. Skilful leadership is particularly important to minority coalitions, such as the Greens, which are usually outgunned on most other resources and must be particularly adroit with the few resources they have.

Measuring leadership and its effectiveness is notoriously difficult, and there is a vast literature on political leadership – much of which is vague or contradictory – attesting to this view. In an effort to avoid becoming mired in this issue, the approach taken here was simple, if not particularly rigorous: interviewees for this study were asked to identify any leaders they thought were particularly important in water policy development. The answers were surprisingly consistent, both across coalitions and across governmental and non-governmental actors. Most pointed to the leadership of Martha Kostuch as a very important intangible factor in the successful policy influence of the Greens, a finding also supported by Glenn (1999).

Kostuch was born in rural Minnesota, trained as a veterinarian, and moved to Rocky Mountain House to set up practice in 1975. By the late 1980s, when she became involved in Friends of the Oldman River, she was already a seasoned veteran of Green political activism. In the late 1970s, she led a successful campaign to reduce sulphur dioxide emissions from the sour gas industry (which were causing immunological and reproductive problems in local cattle), and she organized resistance to the construction of a proposed golf resort on the Cline River (which threatened local aquatic habitats) (Brooymans 2008). In 1987, Kostuch was invited to the inaugural meeting of Friends of the Oldman River and was immediately elected as vice-president, a position she held for most of its existence. Though she was never the organization's leader on paper, she was the driving force behind it, and her work can be credited with much of Friends of the Oldman River's success:

> From the beginning, [FOR's membership] gave Kostuch carte blanche to act on FOR's behalf, with the understanding that she would seek approval from the executive for any deviation from the approved action plan. She was, de facto, chief operating officer for the society. While she frequently consulted [FOR President Cliff] Wallis, and he shared some of the work, Kostuch operated largely on her own, particularly in handling the numerous legal actions in which FOR became involved. No one who has been involved in a volunteer organization will doubt that this arrangement had a lot to do with whatever success FOR achieved. (Glenn 1999, 146)

Kostuch also had a reputation as a fierce and crafty political operative. She became a talisman for the Greens and earned a great deal of respect from the Aggies. Upon her death in 2008, one of her close friends recalled, "There aren't many people in Canada of whom you could say something like, 'We'll sic Martha on you,' and it instils the fear of God in them ... With Martha, you didn't even need to use a last name" (Brooymans 2008).

Highlighting the considerable influence of Martha Kostuch is not to suggest that other leaders in the Green or Aggie coalitions were unimportant. Both coalitions had dedicated and talented leaders who contributed greatly to their respective coalitions' advocacy efforts. However, given the reverence afforded to Kostuch by both participants and informed observers, Kostuch was in a leadership class by herself. For the Greens, this charismatic leadership was an important soft power resource, giving them a focus and enthusiasm they would not otherwise have had.

Explaining the Shift to Softer Policy: The Next Piece of the Puzzle

This chapter has provided the next piece of the puzzle in our explanation of the shift to softer water policy in southern Alberta. Having established the prevailing advocacy coalitions in the previous chapter, this chapter has systematically examined the resources available to each coalition during the period of analysis. The resources were categorized as either hard power or soft power resources in an effort to determine whether there was a shift in the balance of either type of power that might help to explain Alberta's introduction of softer water policy reforms starting in the mid-1990s.

The balance of hard power in the water policy subsystem remained decisively tilted in the Aggies' favour throughout the period of analysis. The Aggies' influence over the irrigation belt ridings, the tendency of single-member plurality electoral systems to over-reward such geographically concentrated interests, the perennial practice of over-representing rural ridings in the Alberta legislature through seat distribution, and the irrigators' tendency to vote for the PC governing dynasty all combined to ensure that the Aggies were well represented in the legislature, the governing caucus, and usually the Cabinet. In contrast, the Greens have had no electoral success in the province and never enjoyed formal legal authority over policymaking in the way the Aggies have. Though the Aggies' hard power has declined in an absolute sense, as a result of their declining share of the provincial population and legislative seats, they still retained far more hard power than the Greens, and the Greens have never come close to displacing them as the dominant coalition in the water policy subsystem. Consequently, Alberta's move to softer water policy cannot be explained as resulting from a shift in the balance of hard power.

When it comes to soft power, the story is considerably different. For decades, the Aggies' claims about the economic promise of irrigation and its importance as a pillar of rural communities went virtually unchallenged. This ended with the emergence of the Greens in the late 1980s and, over time, the Greens cultivated and accumulated important sources of soft power that changed the terms of the water policy debate. Study after study lent credence to the Greens' claims that southern Alberta's rivers were in trouble and that urgent action was needed to protect and restore them. Green policy positions were also supported – sometimes overwhelmingly – by public opinion, and this, in conjunction with the Greens' charismatic leadership and large numbers of mobilizable troops, allowed them to pressure and persuade policy decision-makers, and sometimes even opposing Aggies, into accepting the inherent "rightness" of some of their policy positions. The Aggies tried to counter Green soft power by outspending them on advocacy activities, but these efforts were overwhelmed by the second wave of environmentalism and by urban Albertans' increasing disconnection with agriculture. The overall result was a tilt in the balance of soft power towards the Greens, resulting in a subsystem in which the Aggies relied mostly on hard power and the Greens relied mostly on soft power.

These findings need to be considered in light of the (reformulated) second ACF hypothesis on policy change. This hypothesis states that a shift in the balance of coalition hard power, a shift in the balance of coalition soft power, imposition by a higher-level government, or some combination of these are necessary factors in major policy change. The third factor, imposition by a higher-level government, was ruled out in chapter 3, since the federal government has shown no inclination to involve itself in Alberta's domestic water policy. The findings in this chapter also rule out the first factor but support the second factor: there is no evidence of a shift in hard power, but good evidence of a shift in soft power. This suggests that it was a shift in soft power towards the Greens that was a necessary factor in Alberta's adoption of softer water policy reforms.

This provides the first part of an ACF explanation of Alberta's water reforms, but, as outlined in chapter 2, a full ACF explanation goes much further. The ACF conceives policy development as a longitudinal process unfolding over time; as such, it is shaped by the contingencies of historical events and by policy elites' capacities to learn. These factors are captured in the first ACF policy change hypothesis, which recognizes four different paths of policy reform: a path initiated by an external shock to a policy subsystem, a path initiated by an internal shock to a policy subsystem, a path characterized by policy-oriented learning, and a path involving negotiated agreement between advocacy coalitions. Using the evidence garnered in chapters 4 and 5 as baseline data, the task in chapters 6, 7, and 8 is to identify the "paths" of policy reform evident in southern Alberta water policy. This is accomplished through detailed process tracing of policy development and a comparison of policy development before the Water Act reforms (in chapter 6), during the Water Act reforms (in chapter 7), and after the Water Act reforms (in chapter 8).

Chapter Six

Water Management for Irrigation Use

Many parts of Western Canada experience water shortages from time to time. In response to this problem, governments have invested heavily in water storage facilities, but forecasts suggest that some river basins will be unable to meet future demand for water even when the available storage sites are fully developed ... When potential solutions of this nature are discussed, attention is paid only rarely to the question of whether water rights legislation exacerbates water shortages and, if it does, whether there are any readily available methods of reform.

David R. Percy, professor of law,
University of Alberta

The summer of 1984 was hot and dry in southern Alberta, one of the driest since 1916. Even worse, it was the eighth consecutive drier-than-normal year, and water was beginning to run short. Dryland farmers watched helplessly as their crops withered and died in the fields, and even irrigators became concerned that enough water might not be available to sustain their crops. River flows were at near-record lows and water storage levels in a number of irrigation districts ran dangerously low. More water was desperately needed, but yet the sky refused to rain (Glenn 1999, 40–1).

The water shortage was particularly galling for irrigators in the Lethbridge area, because they believed the worst effects of the drought would have been mitigated with the construction of a dam on the Oldman River, a project that had been first promised by the Lougheed government in 1975. Since then, the dam had undergone numerous rounds of study and evaluation, but a dam site had not been selected and construction had not yet begun. As temperatures peaked in July,

Aggies in the Picture Butte area lost patience with the government's foot-dragging and organized a meeting to express their frustration. On 16 July, six MLAs from the irrigation belt, including Agriculture Minister Fjordbotten, were invited to a local school auditorium where they were confronted by about five hundred angry Aggies who demanded construction of the long-promised dam. The exchanges were heated and left a major impression on the politicians who brought the Aggies' concerns back to Edmonton (Glenn 1999, 40–1).

Less than a month later, Premier Lougheed, Environment Minister Bradley, and a number of local MLAs toured the drought-stricken area and made a series of policy announcements. One was an emergency drought assistance package worth around $30 million. Another was a commitment to begin construction of a dam on the Oldman River no later than 1986. Moreover, the Three Rivers site was selected as the location of the dam, the site preferred by the Aggies but opposed by the local Peigan First Nation, whose traditional lands would be inundated by the dam's reservoir (Glenn 1999, 40–1).

The siting decision for the Oldman Dam is just one example among many of the Aggies' hard power at work in southern Alberta water policymaking. This chapter examines the period between 1971 and 1992, in which the PC Party became Alberta's latest governing dynasty and in which the Aggies dominated the southern Alberta water policy subsystem. It examines the impact of the PCs' election on the policy subsystem and how, over time, the PCs displaced Social Credit as the political partners in the Aggie coalition. The chapter then turns to water policymaking during this period, including the Water Management for Irrigation Use policy, the construction of the Oldman Dam, and the SSRB Water Management Policy. Throughout, the common refrain is the importance of the Aggies' hard power in allowing them to see their policy core beliefs reflected in provincial water policy. The result, at least until the early 1990s, was a predominantly hard approach to water governance.

A Shock to the Subsystem: Enter the Tories

When the PCs came to power in Alberta under Peter Lougheed in 1971, the change in government was – in ACF terms – an external shock to the southern Alberta water policy subsystem. The election was shocking because it disrupted the close relationship that had developed between irrigators and the Social Credit Party, a relationship that was an important aspect of the Aggies' hard power at that time. By 1971, Social Credit had been in power for thirty-six consecutive years and its ties with the irrigation community ran deep. Irrigators were an important part of Social Credit's electoral base, as the irrigation belt consistently elected Social Credit MLAs. Through organizations such as the Irrigation Council and the party's irrigation caucus committee, these irrigation belt MLAs worked closely with the irrigation districts and officials from the Department of Agriculture, forging tight working relationships and building an Aggie coalition that dominated the water policy subsystem. As a result, Social Credit had presided over a massive expansion of irrigation during its tenure, a policy direction it showed every intention of continuing. Party leader Harry Strom represented the southern Alberta riding of Cypress, was a former agriculture minister, and was a big irrigation booster. In 1969, his government dedicated $600,000 per year to the rehabilitation and expansion of irrigation infrastructure, a figure that was later increased to $1 million in 1971. As long as Social Credit remained in power, the Aggies' lock on the water policy subsystem seemed assured.

The election of the Lougheed government was especially shocking to the Aggies because they had played almost no part in it. In 1971, Lougheed's electoral supporters were primarily young and urban, reflecting the demographic changes sweeping Alberta and the rising political power of the cities (Elton and Goddard 1979). In the irrigation belt, however, only one seat went to the PCs, the rest going to Social Credit. For the first time in memory, the irrigators were not well-represented within the governing party, and their hard power within the water policy subsystem was threatened. Accordingly, the PCs' election had shaken the policy subsystem enough that a potential path for policy change had suddenly opened up. Yet this path quickly fizzled out and no change in water policy direction was forthcoming: Alberta continued to follow the hard policy approach it had been following for decades. This was an instance where an external shock failed to stimulate major policy change, and the reasons for this were manifold.

One reason was the absence of an opposing advocacy coalition that might have taken advantage of the external shock and pushed for water policy change. In 1971, there were only two Green groups active in southern Alberta, the Alberta Fish and Game Association and the Federation of Alberta Naturalists, the latter barely a year old. Moreover, the analysis in chapter 4 showed that it was not until the late 1980s that the Green groups actually coalesced into an advocacy coalition with any kind of concerted political power. Nor was there an urban or industrial-based advocacy coalition, which has been an important actor in some water-scarce jurisdictions, but has been a nonentity in southern Alberta. Simply put, when the PCs came to power, the Aggies had the water policy subsystem to themselves, so the external shock wrought by the election put the Aggies on their heels, but there was no opposing advocacy coalition to exploit this opportunity, so no policy change resulted.

Another reason was the decisiveness of the PCs in moving to displace Social Credit as the irrigators' preferred political partner in the Aggie coalition. The irrigation belt did not support the PCs in the 1971 election, but, rather than write them off as implacable political opponents, they redoubled their efforts to court the irrigators and bring them into their electoral base. The PCs quickly built bridges to the irrigation community upon assuming office and undertook a number of Aggie-friendly water policy initiatives between the 1971 and 1975 elections. Moreover, as the PCs courted the irrigators, Social Credit drifted further away from them. Harry Strom resigned as party leader in 1973 and was replaced by Werner Schmidt, an urban intellectual with no ties to the irrigation community. Schmidt was also a party outsider, and his selection virtually split the party, putting Social Credit into a downward political spiral from which it never recovered (Elton and Goddard 1979, 61–2). By the 1975 election, the PC Party's courting and the Social Credit Party's disarray prompted the irrigation belt to swing massively to the PCs, effectively making them the new political partners in the Aggie coalition.

The final reason why the external shock of the 1971 election did not result in significant policy change was that, even though it introduced a new political partner into the Aggie coalition, it did not bring with it a new water governance philosophy. The PCs maintained the same hard water governance philosophy that had pervaded Alberta for decades and, if anything, extended it even further. Peter Lougheed brought with him a more modern and activist approach to government than had existed under Social Credit, and his overriding priority was to use the windfall from Alberta's oil royalties to stimulate economic diversification in the province and achieve a more sustainable prosperity. This was also a time when irrigation sprinkler technology became available and affordable for southern Alberta's irrigators, vastly increasing the amount of potentially irrigable land in the region, and increasing the amount of irrigation water demand. When Lougheed's governance approach was married to the economic potential of increased sprinkler irrigation, the result was a hard water governance philosophy centred on finding new supplies of irrigation water through infrastructure expansion and refurbishment. This approach was encouraged and embraced by irrigators and irrigation bureaucrats, and, under the PCs, hard water policy was not only continued but expanded.

Ultimately, the 1971 election was a shock to the water policy subsystem, but it simply resulted in the displacement of Social Credit by the PCs as the preferred political partner in the Aggie coalition. This was a major change to be sure, but it did not challenge the prevailing hard approach to water policy. Though the shock initially seemed to threaten the Aggies' hard power in the policy subsystem, this power was quickly reasserted as the PCs and the irrigators formed a new political partnership starting in the 1975 election, and the water policies produced between 1971 and 1992 clearly reflect this change.

Until the Water Act reforms of 1996, the focus of water policy in southern Alberta was supply management. Essentially, government saw its role as providing water supplies when and where they were needed, to meet growing water demands. To do this, priority was placed on the refurbishment and expansion of water management infrastructure, bringing more and more water under the control of government water managers, who could then distribute this water among the growing ranks of water licence holders, as demands warranted. The network of water management infrastructure constructed during this period was both large and comprehensive, ranging from the construction of large dams on the Red Deer, Paddle, and Oldman Rivers, to the refurbishment of small irrigation district canals through a perennial program of government subsidies. It is important to note, however, that these supply management efforts were restricted to intra-basin engineering efforts only. For reasons explained below, the Lougheed government decided at an early stage that large inter-basin diversions, such as PRIME, were a non-starter for them. So even though they pursued an overwhelmingly hard water policy, the scope of this policy was limited to within-basin supply management.

In the remainder of this chapter we investigate the hard water policies of 1971 to 1992, exploring the politics behind them and the role of the powerful Aggie coalition in their adoption. We start with an examination of the early Lougheed governments and their adoption of the Water Management for Irrigation Use policy, which established a water policy blueprint for the next twenty years. Then we will examine the two most significant policy initiatives of the period, the building of the Oldman River Dam and the introduction of the South Saskatchewan River Basin Regulation, the former a very important supply-side measure, and the latter an important demand-side measure.

Water Management for Irrigation Use

The time between the elections of 1971 and 1975 can best be described as the courting period between the PCs and the irrigators. Two months after assuming office, and despite having only one member from the irrigation belt, a meeting was arranged between the new government and representatives from the Alberta Irrigation Projects Association. Attending on the government side were no fewer than six members of Cabinet, including the ministers of agriculture, highways and transportation, municipal affairs, environment, federal and intergovernmental affairs, and lands and forests (Dowling to Thiessen, 13 October 1971). The government was keen to show irrigators that it was open to their concerns, and the meeting made a favourable impression on them, as reported by the head of the Irrigation Projects Association in his yearly address to the membership: "After our recent discussions in Edmonton with several Cabinet ministers, I can say they were very warm and friendly. We presented several problems which were well received and our feelings were quite mutual. They assured us they had our interests at heart and would do all they could to help us at both the Provincial and Federal levels of Government" (Noble 1971).

In the years that followed, the new government made good on this commitment, pursuing water policy initiatives in close consultation with irrigators and heavily informed by irrigators' policy preferences. One early indication was the government's stance on water pricing.

In December 1972, the Planning Division of Alberta Environment produced a policy document entitled "General Principles Governing Water Management in Alberta." At this time, the PC government and the department itself were both just over a year old, and the document was intended as a general statement of water policy, spelling out how the new administration intended to approach water governance. Much of the document was an endorsement of the status quo, including government ownership and licensing of water use, compliance with the Master Agreement on Apportionment, water management on a river basin basis, and consideration of environmental impacts in all water development projects. The only significant departure was its speculative endorsement of water pricing as a policy instrument for incentivizing increased conservation and efficiency of water use among the province's water users, particularly irrigators (Alberta Environment 1972).

The general principles document was sent by departmental staff to Environment Minister Yurko for approval and release, and that is where it stopped. It was never publicly released and never endorsed as a statement of government policy. According to a government official interviewed by de Loë, the document was halted because it was not considered "politically or administratively palatable" (de Loë 1994, 210). More specifically, the irrigators had made it very clear in their early meetings with government officials that they would not accept any "taxes" (in their words) on their water use; so, to avoid alienating the irrigators, the government avoided any official policy statements that might imply its endorsement of the use of water pricing. A few months later, the government made their non-use of water pricing official when Agriculture Minister Horner stood in the legislature and stated, "The former [Social Credit] government used to fly kites all the time and say, well, we're going to charge you more for your water. We'll have to put a user fee on. We want to clear that up pretty quickly, Mr Speaker. We don't intend, other than the normal fee they pay through their irrigation district, to put any user fee on the water for irrigation" (Legislative Assembly of Alberta 1973, 434).

Needless to say, irrigators warmly welcomed this statement, as it directly reflected their policy preferences. This early policy commitment also had long-lasting effects: the irrigators continued to object to water pricing as a policy instrument right through to the mid-2000s and have successfully defended this position so that increased water consumption fees did not feature in the eventual shift to softer water policy.

Another example of the burgeoning alignment between irrigators and the PC government was in the refurbishment and expansion of water infrastructure in the irrigation districts. At the time, this issue was entangled in federal-provincial politics, considerably complicating it. The federal government had been involved in developing the irrigation districts since the late 1930s, but by the late 1960s was signalling its intent to get out of the irrigation business. Federal-provincial negotiations concerning the federal withdrawal had started under the previous Social Credit government and dragged on through the first two years of the new PC government. At issue was the transfer of federal assets in the Bow River and St Mary irrigation districts, as well as how much Ottawa would contribute to an ongoing infrastructure refurbishment fund. Irrigators, bolstered by the expansion opportunities afforded by sprinkler technology, identified the conclusion of a federal-provincial agreement as a high priority and put considerable pressure on the Alberta government to achieve it. At one point, they even requested a seat at the negotiating table in order to help move things along, a request that was declined by the Alberta government (Noble 1972; H. Horner to R.T. White 28 April 1972).

Finally, in 1973, the Canada-Alberta Irrigation Agreement was concluded. The agreement transferred all federal irrigation assets to the province of Alberta and set up an infrastructure refurbishment fund of $2 million per year, with the federal contribution phasing out over a number of years. The Alberta government also committed itself to running the irrigation headworks[8] acquired from Ottawa at no cost to the districts, a commitment that was later extended to include all irrigation headworks in the province. In other words, the province pledged to absorb the costs of operating and refurbishing district headworks rather than passing them on to the water users, providing the irrigation districts with a generous operating subsidy that had been on their policy wish list for many years.

As the 1975 election approached, the courting of the irrigators by the PC Party intensified. In late December of 1974, with election speculation rampant, Conservative candidates from irrigation belt ridings met with leaders from the Alberta Irrigation Projects Association to discuss water and irrigation issues. For the politicians, the aim was to reassure the irrigation community that the PCs had their best interests at heart and that they were safe to elect a PC government. For the irrigators, the meeting was a chance to extol the virtues of irrigation to prospective policymakers, and they made the most of the opportunity. The politicians left the meeting with a brochure comparing irrigation farming to dryland farming and showing that irrigation resulted in far more productive use of the land: $190 return per acre versus $33 return per acre. Weeks later, some of these MLAs cited directly from this brochure as they made statements in the legislature in support of the government's

8 The headworks are the dams and weirs used to divert water from rivers into the irrigation districts. From this point, the water is distributed throughout the districts through networks of canals.

Water Management for Irrigation Use policy (Legislative Assembly of Alberta, 12 June 1975, 678).

For irrigators, however, the most important part of the 1975 election campaign was the Lougheed government's promise to build a dam on the Oldman River. Irrigators saw this as essential for further irrigation expansion in the upper reaches of the Oldman sub-basin and had been calling for a dam on the mainstem of the Oldman for many years. This promise was the master stroke in the PCs' courting of the irrigators, and the election results are telling of its success.

The 1975 election was a landslide victory for the Lougheed PCs, and this time the irrigation belt was counted among their supporters. The PCs claimed sixty-nine of the seventy-five seats in the Alberta legislature, giving them a much wider margin of victory than in 1971. In the irrigation belt, eleven of thirteen seats went to the PCs as the irrigation community decisively shifted away from Social Credit. Social Credit leader Werner Schmidt ran in the constituency of Taber-Warner – in the heart of the irrigation belt – but lost to a Conservative candidate. This was Schmidt's second attempt to win a seat in the legislature, having already failed in a by-election attempt just over a year previously, and the loss, combined with the party's poor showing, prompted his resignation as party leader. So, while the results of the 1975 election left Social Credit in further disarray, they also solidified the PCs' emergence as the new "natural governing party" of Alberta, drawing support from just about all segments of the electorate, including the irrigation community (Elton and Goddard 1979). The PCs' displacement of Social Credit in the Aggie coalition was now virtually complete, as evidenced by the irrigation belt's consistent election of PC MLAs over the next three decades.

In the wake of the election, the reinvigorated Aggie coalition kicked into action. Just a few days before the opening of the new legislature, the ministers of agriculture, environment, and Native affairs toured the irrigation districts and met with the district boards to discuss expansion and refurbishment of district infrastructure (Legislative Assembly of Alberta, 20 May 1975, 31–2). In the government's subsequent Throne Speech, it committed – somewhat cryptically – to the announcement of new "irrigation policy guidelines" very soon (Legislative Assembly of Alberta, 15 May 1975, 2). Five days later, the government filed a pamphlet in the legislature entitled *Water Management for Irrigation Use*, which outlined a series of policy commitments aimed at "expansion of irrigated areas and rehabilitation of existing irrigation works" (Legislative Assembly of Alberta, 20 May 1975, 26).

Authored jointly by the Departments of Agriculture and Environment, *Water Management for Irrigation Use* outlined a hard water policy blueprint that would be followed in southern Alberta for the next seventeen years. Rather than a new policy direction, it was a consolidation and extension of hard policy measures already in existence. The underlying policy objective was to expand irrigated acreage in southern Alberta by up to 700,000 acres over the next several years (Alberta Environment and Alberta Agriculture 1975, 3). To this end, it reconfirmed the government's promises to construct an Oldman River dam and to take over the operation and funding of the irrigation districts' headworks. This would be further supplemented by a $200 million commitment of funds from the recently created Heritage Savings Trust to support the refurbishment and expansion of infrastructure within the districts. It also committed to making irrigation water use a higher priority under the Water Resources Act, a pledge that was kept through legislative amendments later in the year (1).

The Water Management for Irrigation Use policy, which was both ambitious and expensive, was greeted with almost universal approval when debated in the legislature. A lot of the discussion focused on the $200 million allocated from the Heritage Savings Trust, a fund set up by the Lougheed government to take some of the province's oil revenues and invest them in economic diversification. MLAs applauded the anticipated economic impacts of both the prospective Oldman Dam and the irrigation expansion program. As one MLA stated, "I would like to suggest that the type of thing the heritage trust fund should be used for is water control, water use and management throughout the province. This is something that is discussed throughout North America and is probably ultimately our most valuable resource. The kind of benefits that would accrue – irrigation, controlling levels of lakes, assuring future supplies of water to the southern part of the province – would be a useful way to take advantage of our fortunate position and would benefit future generations" (Legislative Assembly of Alberta, 12 June 1975, 678).

Another MLA suggested that the only shortcoming of the policy was that it had not been extended to the dry areas of east-central Alberta, where a program of dam and irrigation district construction was also needed (Legislative Assembly of Alberta, 12 June 1975, 679–80). This overwhelming support for the Water Management for Irrigation Use policy is hardly surprising, given the government's alignment with the Aggie coalition and the extent to which the government dominated the legislature after the 1975 election. Moreover, two of the four MLAs in the opposition Social Credit caucus hailed from the irrigation belt, so the official opposition was not inclined to oppose irrigation-friendly policies.

With the Water Management for Irrigation Use policy in place, the government had effectively committed itself to "finding" more water in southern Alberta and dedicating this water to irrigation expansion. The centrepiece in the effort to find more water was the construction of a dam on the Oldman River, a fateful decision that would turn into a policy odyssey and become the epitome of the hard policy era.

The Oldman River Dam

While the PCs aligned themselves with the Aggies and shared their hard vision of water governance, there were limits to this shared vision. The new government was forced to address one of these limits in early 1972 after the Prairie Provinces Water Board released its massive Saskatchewan-Nelson Basin Study. This study looked at dozens of undeveloped dams and inter-basin diversions across western Canada and provided preliminary assessments of their viability and potential contributions to the regional economy (Godwin 1981). The study's release once again raised the possibility of constructing a large north-south inter-basin diversion in Alberta, such as the PRIME project, forcing the new government to stake its position on inter-basin diversions.

For the Lougheed government, PRIME was something of a political minefield. On one hand, the project was heavily favoured by irrigators and many other southern residents who saw it as key to future economic development in the south. On the other hand, PRIME promised to be massively expensive, to dislocate hundreds of landowners, and to dramatically affect the environments of central and southern Alberta. Some critics also saw PRIME as a first step in commencing large-scale water exports to the United States, a threat that was probably more imagined than real, but alarmed many Albertans nonetheless. The new government resolved the dilemma by indefinitely shelving PRIME in April 1972, but in a way that still allowed for further large-scale water development in the south. As Environment Minister Yurko explained in the legislature, no inter-basin diversions would be considered by the government until "total water basin development" had been achieved (Legislative Assembly of Alberta 1972). That is, all available measures for improved intra-basin supply management would be undertaken before inter-basin diversions would be contemplated. PRIME was not killed by this decision, but it was made clear that PRIME was not in the government's immediate water policy plans.

In the absence of inter-basin diversions, on-stream dams took on additional importance as instruments of water supply management. Dams were important because they allowed large amounts of surface water to be stored when available (usually in the spring) and released when needed by water users (usually in the summer). This allowed for more complete use of available water resources, at least for human economic purposes. Many dams had already been constructed around southern Alberta over the previous sixty years, but the next generation of dams would be considerably larger than in the past and would bring some of the region's largest rivers under the control of water managers (see table 3.1 for a list of southern Alberta's dams). Between 1971 and 1975, proposals surfaced to construct large dams on the Paddle, Red Deer, and Oldman Rivers, all of which were supported by the Lougheed government. Of these, the Oldman Dam would be by far the largest and would have the greatest impact on the irrigation community.

The decision to build the Oldman Dam was strongly motivated by the PCs' desire to secure irrigators' political and electoral support. Irrigators had wanted a dam on the Oldman River for decades, but in the summer of 1974 they picked up signals from the government that this was becoming a serious possibility. In June, Alberta Environment was directed "to compile and finalize a water management plan within the Oldman River System," a plan that could include a new on-stream dam (W. Solodzuk to J.G. O'Donoghue 27 June 1974). Sensing an opportunity, the Alberta Irrigation Projects Association met with Environment Minister Yurko in July to lobby for a dam and, in conjunction with various local governments in the Oldman sub-basin, organized a petition to show widespread local support for a dam (F.A. Ross to Town of Nobleford 29 August 1974). In effect, the PCs floated the Oldman Dam idea in mid-1974, and, having gotten an enthusiastic response from Aggies, incorporated it into their platform for the 1975 election. The official announcement was saved for the campaign trail and was one of the key factors in swinging the irrigation belt from Social Credit to the PCs.

The campaign promise was subsequently incorporated into the Water Management for Irrigation Use policy, and attention was soon turned to the "technical" aspects of the dam, such as dam size, siting, and design. Between 1974 and 1978, two rounds of study were undertaken by Aggie-dominated committees to address these issues. The first round (1974–6) was a continuation of the Oldman River planning commenced by the Department of the Environment in mid-1974 and involved a committee consisting of representatives from the provincial Agriculture and Environment Departments, the federal Prairie Farm Rehabilitation Administration, local governments, irrigators, and businesses. The second round (1977–8) was undertaken by a group known as the Oldman River Basin Study Management Committee, comprising four senior provincial public servants, three ranchers, a dryland farmer, an irrigation farmer, and a business executive. There were no Greens on either committee and only the second one considered any water management alternatives to damming the Oldman River. While the first committee recommended that a dam be constructed at the Three Rivers site, the second recommended either the Three Rivers or Brocket sites, leaving the eventual siting decision in the government's hands.[9]

Dam siting turned out to be a thorny issue for the government, not only on the Oldman River but in other places as well. In 1973, the Alberta government decided to build a dam, now known as Dickson Dam, on the upper reaches of the Red Deer River. This decision was met with considerable opposition from local landowners whose lands would be inundated by the dam and who tried unsuccessfully to reverse the government's decision. Similarly, in the wake of the 1976 report, local landowners formed the Committee for the Preservation of Three Rivers to resist any dam-building plans there. The Three Rivers site was further complicated by the fact that it was adjacent to traditional Peigan First Nations lands that were likely to be inundated, arousing opposition from the Peigan. Nevertheless, the Three Rivers site was favoured by most Aggies, and it was the site endorsed by both study reports, so it remained high on the government's shortlist.

By the late 1970s, the Oldman Dam was also facing criticism from environmentalists. They pointed out that neither round of study had adequately considered the dam's environmental impacts, which, they argued, would be quite severe: native fish stocks, wetlands, and riparian forests all promised to be negatively affected by the dam's construction. In an attempt to manage these criticisms, the government referred the project to the Environment Council of Alberta for review in late 1978.

9 The Three Rivers site was so named because it was located near the confluence of the Oldman River with two smaller rivers, due north of the town of Pincher Creek. The Brocket site was located further downstream on the Oldman River near the small town of Brocket and on the reserve lands of the Peigan First Nation.

The council was an arm's-length public body with a mandate "to review government policies and programs and inquire into matters pertaining to environment conservation" (Provincial Archives of Alberta 2006, 190). Notably, the referral was made only months before the 1979 general election, temporarily defusing the Oldman Dam issue and helping the PCs win another landslide election victory.

The Environment Council's Oldman review committee comprised Arnold Platt (an agrologist and farmer), Allistair Crerar (the Environment Council's executive officer), Thomas Sissons (a businessman from Medicine Hat), and Dixon Thompson (from the Faculty of Environmental Design at the University of Calgary). Its mandate was to "enquire into the conservation, management and utilization of water resources within the Oldman River Basin … [as well as] the merits of alternative means of providing for future water requirements" (Environment Council of Alberta 1979, 4). It did so by inviting public submissions and conducting public hearings at eight locations around southern Alberta, hearing the views of all stakeholders who cared to get involved (6). (The submissions and transcripts from these hearings were sampled and analysed as part of the content analysis in chapter 4). As shown in chapter 4, both Greens and Aggies were active in the consultations, though the Aggies were far more numerous and much better organized. Both sides took predictable positions on the Oldman Dam issue, with the Greens consistently against and the Aggies consistently in favour of dam construction. The analysis in chapter 4 also showed how deeply rooted the hard policy approach was at that time, even among many Greens who argued against the dam by suggesting off-stream storage and infrastructure refurbishment projects as acceptable alternatives.

The Environment Council released its report in August 1979 and, remarkably, recommended against the construction of a dam on the Oldman River. As stated in the final report, "The Environment Council does not believe that an onstream dam is required to provide for the development of irrigation to its maximum economic potential in the basin" (Environment Council of Alberta 1979, 145). Instead, it recommended an accelerated program of infrastructure refurbishment and the construction of off-stream storage at various points in the Oldman system. It also recommended a number of soft policy measures, such as the adoption of minimum streamflow levels and increases in water pricing to incentivize water use conservation and efficiency (viii). This combination of hard and soft policy measures, the council argued, would allow for continued irrigation growth and would make the construction of an on-stream dam entirely unnecessary.

Not liking the advice it had received from the Environment Council, the government created another ad hoc water policy advisory body in April 1980 and stacked its membership with card-carrying Aggies. Known commonly as the Kroeger Committee, after its chair, Transportation Minister Henry Kroeger, it was asked "to advise on the need for new policy in respect to long-term water resources planning and management in relation to balanced economic development in the province." The committee comprised sixteen members, including two retired judges, two farmers, two engineers, two businessmen, one university professor, five public servants, and three MLAs, two of which were farmers. Kroeger himself was a farmer who represented the arid constituency of Chinook in southeastern Alberta and was a well-known irrigation booster. A few years later, in fact, he would be appointed to the provincial Agricultural Hall of Fame for his work in irrigation advocacy. As one observer noted at the time, "The makeup of this committee hints at a structural [i.e., hard] approach to the problem, a preoccupation with meeting demands and little or no indication that significant and adequate consideration will be given to environmental and ecological concerns and to demand management and conservation as important management tools" (Thompson 1981, 104). After a year of deliberation, the Kroeger Committee reported to Cabinet in the summer of 1981, endorsing an expansion of the hard policy approach in southern Alberta, including the construction of an Oldman Dam and even the building of an inter-basin diversion to bring water down from the north (Glenn 1999, 43). These recommendations helped to counterbalance the Environment Council's earlier recommendations, giving the Aggies further justification for continuing with their dam.

In some respects, though, the Kroeger Committee was rendered irrelevant by events before it even reported. In August 1980, a year before the Kroeger Committee reported, the Alberta government announced that a dam would be built at either the Three Rivers or Brocket sites, pending the outcome of negotiations with the Peigan First Nation. These negotiations concerned the enlargement of the headworks for the Lethbridge Northern Irrigation District. Lethbridge Northern would be the primary beneficiary of the dam, wherever it was located, and the district's headworks were on Peigan reserve land. These headworks needed to be expanded in order to take advantage of the additional water provided by the new dam, and this necessarily involved the surrender of some reserve land to the provincial government. After eighteen months of negotiations, a deal was reached between the Peigan and the government in April 1981, and one of the final hurdles for dam construction was overcome (Glenn 1999, 38–40). All that remained was to select a dam site, which still remained a contentious issue.

Matters came to a head in the summer of 1984 when, as described in the opening to this chapter, pressure from the Aggies finally forced the government's hand. Since 1977, southern Alberta had experienced a series of drier than normal years, with 1984 being the driest since 1916. During this time, administrative moratoria were imposed on new water licences in the upper reaches of the Oldman sub-basin due to the general water shortage, and a number of irrigation districts, including Lethbridge Northern, struggled to meet the water orders of their irrigators. The ongoing drought prompted the public meeting between MLAs and irrigators at Picture Butte on 16 July, where the irrigators expressed, in no uncertain terms, their displeasure with the delays in dam construction. "Whatever message the beleaguered MLAs took back to Edmonton, whether concern for the plight of the farmers or concern for the future of the Conservative Party, it had an obvious impression on Premier Peter Lougheed" (Glenn 1999, 40–1). Less than a month later, Lougheed visited Lethbridge to see the drought devastation first-hand and to announce publicly that construction of a dam at the Three Rivers site would begin in 1986, eleven years after initially promised.

While the dam had always had critics, a concerted opposition to the dam began to form over the next three years, much of it based on environmental grounds. In the wake of Premier Lougheed's announcement, the Committee for the Preservation of Three Rivers renewed its efforts to resist the dam, but by mid-1987 the government had acquired all of the land necessary for construction, and the concerns of local landowners either dissipated or were absorbed into the environmentalist movement (Glenn 1999, 48). The Peigan, who strongly favoured the Brocket site, also organized to resist the Three Rivers dam, often working on their own and sometimes in conjunction with environmentalists. The biggest change, though, was the emergence of a concerted Green advocacy coalition with the formation of Friends of the Oldman River in August 1987. Whereas prior Green opposition had been disjointed and sporadic, Friends of the Oldman River created a coordinative umbrella for Green advocacy efforts. It was also a key factor prompting the Aggies to organize the Southern Alberta Water Management Committee, a pro-dam umbrella organization, only four months later. The Aggies now found themselves in the novel position of being opposed at every turn, and the water policy subsystem quickly took on a dynamic of Aggie-Green competition, each coalition working to outmanoeuvre the other in influencing policy decision-makers.

The Greens and the Peigan presented determined opposition to the dam, but it was very much "outsider" opposition compared to the "insider" status of the Aggies. The Greens had emerged as a force in the policy subsystem, but they had little access to the inner processes of water policymaking. Most Aggies and many neutrals within government regarded the Greens as a fringe protest movement, likely to fade away as environmentalism went out of fashion. Many also regarded the Greens as an illegitimate interest, because most Greens did not have a material stake in water governance in the same way that irrigators did. Consequently, the government did not go out of its way to consult Greens on water policy decisions, in contrast to the Aggies, who were regularly consulted. Moreover, the Greens did not have MLAs representing their cause within the governing caucus and Cabinet as the Aggies had. This left the Greens on the outside of policy development processes looking in, and they resorted to the same tactics that many other political outsiders do: public education campaigns, media campaigns, public demonstrations, and court action.

One aspect of the Greens' strategy was to cultivate favourable public and media opinion, using this as a soft power resource to pressure the government into halting the dam. Friends of the Oldman River, and particularly Vice-President Martha Kostuch, proved very adept at drawing public attention to the Oldman Dam and keeping it in the news. Analysis by de Loë shows that coverage of the Oldman Dam in Edmonton and Calgary daily newspapers spiked in 1987, the same year Friends of the Oldman River was formed, and remained much higher than pre-1987 levels through the early 1990s (de Loë 1999, 223). Part of this had to do with the Greens' tireless advocacy efforts, which continually drew the media to them, but part of it was also Kostuch's charismatic personality and her talent for producing snappy quotes and sound bites. The Greens also engaged the Aggies in the local and provincial media, writing opinion pieces and letters to the editor to rebut and discredit Aggie claims about the benefits of the dam and its importance to the future of southern Alberta.

A number of public demonstrations against the dam were also organized, at the damsite, at government offices in Lethbridge, and at the legislature in Edmonton. The biggest of these events was in June 1989 when Friends of the Oldman River held an outdoor benefit concert, headlined by folk singers Gordon Lightfoot and Ian and Sylvia Tyson and with a special appearance by environmentalist David Suzuki. This single event attracted well over ten thousand people and raised over $20,000 in donations (Glenn 1999, 66). The Greens also gained a measure of public and media respect by disavowing the use of violence: when a First Nations group known as the Lonefighters tried to resist the dam's construction through force of arms, they were denounced by Friends of the Oldman River, who were firmly committed to resistance through only peaceful political and legal means. The larger socio-cultural context also helped the Greens. The Oldman Dam controversy coincided with the "second wave" of environmentalism in Canada, making local and national populations particularly receptive to the Greens' message (Paehlke 1992).

Over time, the Greens' public and media advocacy campaigns paid dividends. As de Loë shows, two of Alberta's major daily newspapers, the *Calgary Herald* and the *Edmonton Journal*, were quite critical of the dam, with negative coverage outstripping positive coverage in most years between 1987 and 1992 (de Loë 1999). Regional media coverage within southern Alberta was more positive, but even some of these newspapers, such as the *Lethbridge Herald*, took editorial positions against the dam (*Lethbridge Herald* 1990). De Loë further argues that this media coverage played an important role in shaping public opinion (de Loë 1999, 235). Opposition politicians picked up on the growing sense of public unease about the dam, particularly in Calgary and Edmonton, and used this ammunition to attack the project during Question Period. All of this put considerable pressure on policy decision-makers, who continually found themselves on the defensive on the Oldman Dam issue.

Another aspect of the Greens' strategy was to block construction of the dam in the courts or, failing that, to undermine the dam's credibility by putting it on public trial. In September 1987, Friends of the Oldman River challenged the legality of the provincial licence issued for the dam, claiming that it had been issued without adequate public notice, contrary to the requirements of the Water Resources Act. The Alberta Court of Queen's Bench sided with the Greens and the licence was quashed. This led to an emergency debate in the legislature, the filing of a government appeal, and the issuing of a new licence, which was subsequently upheld by the courts. Having exhausted this legal route, Martha Kostuch then laid charges against Alberta's environment minister under the federal Fisheries Act for destruction of fish habitat in the Oldman River. Eventually, the case was taken over by the provincial attorney general and the charges were dropped, ostensibly due to lack of evidence. Through all of these legal actions, Friends of the Oldman River had also been working to have Ottawa apply its environmental assessment process to the dam, arguing that federal approval was required according to its Environmental Assessment and Review Process Guidelines Order. When Ottawa declined to proceed with an assessment, the Greens went to court to force them. After a number of rulings and appeals, the dispute went all the way to the Supreme Court of Canada, which, in January of 1992, ruled in the Greens' favour and forced a federal assessment (Glenn 1999).

The report of the Oldman River Dam Environmental Assessment Panel in May 1992 was a high-profile and expert-based vindication of the Greens' position on the Oldman Dam. The panel concluded that "the environmental, social and economic costs of the project are not balanced by corresponding benefits and ... that, as presently configured, the project is unacceptable" (Oldman River Dam Environmental Assessment Panel 1992, 6). The panel's first and "preferred" recommendation was to decommission the dam and restore natural river flows; but, recognizing that this was unlikely, it also provided a secondary set of recommendations to mitigate the dam's adverse environmental effects once it came into operation (6). The report even went beyond the dam itself to criticize the Alberta government's hard policy approach, in general. To deal with water scarcity issues in the future, the report recommended that Alberta avoid building more "environmentally damaging projects" and instead give greater emphasis to softer policy measures such as water pricing and the establishment of minimum instream flows (12). For the Greens, the report was a significant moral and public relations victory, reflecting and contributing to their growing soft power in the Alberta water policy subsystem.

Through all of the Greens' mounting pressure, however, the Alberta government stood by the Oldman Dam, primarily in response to the Aggies' hard power. From the mid-1980s through the early 1990s, the irrigation belt became an even more important part of the PCs' electoral base as, under the leadership of Don Getty and then Ralph Klein, the party lost its grip on seat-rich Edmonton. The 1986, 1989, and 1993 elections saw the PCs consistently sweep the irrigation belt, putting plenty of Aggies within the governing caucus, some of them in Cabinet positions. Moreover, the irrigation belt MLAs and the ministers of environment and agriculture met regularly with the Southern Alberta Water Management Committee to discuss dam-related issues, a privilege not enjoyed by Friends of the Oldman River. The Aggies mounted their own media and public relations campaign in support of the dam, but outside of those who already supported them in southern Alberta, these efforts seemed to have little effect. Instead, it was their hard power that won the day and influenced provincial policy decision-makers to see the dam through to completion.

By the time the Oldman River Dam Assessment Panel issued its report, the dam was almost fully operational and the Alberta government had no intention of decommissioning it. The panel's report was disregarded by the Alberta government, who maintained the position that they did not need federal approval for a provincial project (Glenn 1999, 113). Unwilling to provoke another resource conflict with Alberta, particularly in the midst of the Charlottetown Accord constitutional reform efforts, Ottawa also largely ignored the report. Friends of the Oldman River went back to court to try to force the federal government to act on the recommendations, but the court ruled that Ottawa had only to act on those recommendations with which it agreed, and it chose not to act on any of them (117–19). This left the matter entirely to the province, where the Aggies were the dominant advocacy coalition shaping water policy, effectively sealing a decision in favour of the dam.

Over twenty years later, the Oldman Dam still stands, a monument of progress for the Aggies and a blight on the landscape for the Greens. Aggie-Green skirmishes over the dam's operation still take place, particularly on the Oldman River Dam Environmental Advisory Committee, a multi-stakeholder body established to advise Alberta Environment on the operation of the dam and the mitigation of its environmental impacts. The dam itself, though, is a fact of life, as is the irrigation expansion that accompanied it, the last significant irrigation expansion in southern Alberta. However, construction of the dam did not automatically dedicate the newfound water to support irrigation; this was the result of a separate but related water policy process, one culminating in the SSRB Water Management Policy of 1990.

The SSRB Water Management Policy

The 1975 Water Management for Irrigation Use policy made it clear that irrigation was a high-priority water use in southern Alberta, and that the additional water made available through dam construction and infrastructure refurbishment was intended to support irrigation expansion. As stated in the policy, "Alberta Environment's strategy is to identify most of the excess water throughout the southern part of the Province for agricultural use on a priority basis" (Alberta Environment and Alberta Agriculture 1975, 1). The Alberta government, at the urging of the Aggie coalition, would follow this strategy for the next seventeen years, first defining irrigation as a high-priority water use through legislative amendments and policy statements, and then allocating more water for irrigation through basin-level water planning programs. This recognition as a high-priority use became increasingly important as water resources became scarcer and irrigators faced competing demands from other water users, including the natural environment.

The influence of the Aggies in shaping the Water Management for Irrigation Use policy was described above, and one of the interesting features of this policy was a commitment to amend the Water Resources Act to promote irrigation ahead of other economic water uses in the province. For decades, the Water Resources Act had contained a list of preferential water uses in which irrigation was listed fourth, behind domestic, municipal, and industrial water uses, respectively. The purpose of this list was to allow for the expropriation of licences from inferior to superior water uses, subject to approval by the minister of environment (Percy 1988, 25). Shortly after the 1975 election, amendments were introduced to promote irrigation from fourth to third in the statutory list of preferred water uses, behind only domestic and municipal water uses, which, as basic human consumptive uses, were never likely to be supplanted. Since the list of preferential uses was rarely used, the promotion of irrigation ahead of industrial uses was more symbolic than substantive, but irrigation was now legislated as the highest-priority economic water use in the province, a clear reflection of the Aggies' dominance of the Alberta water policy subsystem (25).

The new prioritization of water uses was reiterated three years later when the Alberta government released a new policy document entitled *Water Resources Management Principles for Alberta*. Despite its lofty title, this document was not a major water policy initiative; it was so low-key that it was not debated in the legislature and does not turn up anywhere in the Hansard record. The *Water Resources Management Principles* were mostly a restatement of the commitments of the Water Management for Irrigation Use policy, as evidenced in its statement on priority water uses: "As basic water management policy, the Alberta Government recognizes the importance of water for human consumption, for food production, for industrial use, and for other uses in that order. This order of preference is embodied in the Water Resources Act" (Alberta Environment 1978, 10). It also restated the government's goal of "assuring a secure and continuous water supply to the Irrigation Districts" and reiterated its pledge that "the waters in each major basin must be fully and efficiently utilized before inter-basin augmentation could be considered" (10–11). In order to achieve full and efficient water use in each basin, and the broader goal of "multi-purpose use," the government committed to managing water on a river basin basis and identified the SSRB as its first priority (10).

As a first step towards basin-level management, a major water planning process was launched in 1980. Known as the South Saskatchewan River Basin Planning Program, it was designed to "identify and prioritize water use objectives for the basin and to determine the most appropriate strategy for water management to achieve those objectives" (Alberta Water Resources Commission 1986, 1). Essentially, Alberta Environment's Planning Division undertook a major study of the water resources, uses, and needs in the SSRB, then developed water quantity and quality models to illustrate the effects of various planning options. A report detailing these results was released to the public in August 1984, and public hearings on the report were held at ten locations throughout the SSRB a few months later. A summary of the public input was itself released in 1986, and was meant to inform the water management options eventually selected by policy decision-makers (Alberta Water Resources Commission 1986).

Despite the copious amounts of water use data and public input involved in the South Saskatchewan River Basin Planning Program, it was still heavily skewed towards Aggie water policy preferences. Though the program ostensibly inquired about "water use objectives" for the SSRB, in reality, the pre-existing objective of irrigation expansion was not challenged, and most of the models developed by Alberta Environment incorporated irrigation expansion as a key parameter. This substantially narrowed the scope of inquiry and limited serious discussion to only those objectives that were compatible with irrigation expansion. Moreover, the public hearings were conducted by the Alberta Water Resources Commission, a new Aggie-friendly water policy advisory body created in December 1982. The commission was created as an alternative to the Alberta Environmental Conservation Authority, which had recommended against the construction of the Oldman Dam in 1978 and had become increasingly marginalized by the government in water policy issues. The commission was headed by PC MLA Henry Kroeger of the recently completed Kroeger Committee. With Kroeger at the helm of the public consultations, the government – and the Aggies – could be confident that their irrigation expansion agenda would not be displaced.

Indeed, the Alberta Water Resources Commission's final report, released in October 1986, endorsed irrigation expansion and the hard policy approach; yet it also recommended a number of softer policy measures in conjunction with irrigation expansion, perhaps reflecting the influence of Greens in the public consultations. The consultations were held only three months after the government's announcement that a dam would be constructed at the Three Rivers site, prompting a number of Green opponents to attend the commission's hearings, the only outlet available for them to voice their displeasure to the government. The Alberta Fish and Game Association, Alberta Wilderness Association, Federation of Alberta Naturalists, Trout Unlimited, and a handful of other Greens submitted briefs against the Oldman Dam and calling for softer water policy alternatives. This is perhaps why the commission's report, while endorsing irrigation expansion, the prior allocation system, and the refurbishment and expansion of water infrastructure, also recommended the establishment of "realistic" minimum streamflows, the continuation of administrative licensing moratoria on heavily allocated streams, and the consideration of water pricing as a conservation measure (Alberta Water Resources Commission 1986). These softer measures were recommended as additions to the current hard policy measures, foreshadowing the type of policy sedimentation that would eventually occur with the Water Act reforms in 1996.

After the commission's report, the water policy subsystem became consumed with the Oldman Dam conflict for about the next three years, and the report was effectively shelved. But as the dam neared completion, so did the prospect of having additional water supplies to allocate, and the government returned to the SSRB water allocation issue. In May 1990, Environment Minister Klein announced the SSRB Water Management Policy to "guide the management of the surface water resources throughout the SSRB" (Alberta Environment, *South Saskatchewan River Basin Water Management Plan* 2002, appendix D). The two most novel elements of the policy had to do with irrigation expansion and instream flows. With respect to irrigation expansion, the government committed to determining how much water would be made available to support future irrigation expansion in the Red Deer, Bow, and

Oldman sub-basins. It also committed to developing "preferred" and "minimum" instream flow levels on the regulated rivers of the SSRB, and to operating dams on these rivers so that preferred levels would be met most of the time, dropping only to minimum levels during low run-off periods. The details of both commitments – the actual levels of irrigation expansion and instream flows – would be determined after consultations with relevant stakeholders (ibid.).

The stakeholders the government had in mind were the Aggies, but not the Greens. Over the next sixteen months, the Departments of Environment and Agriculture worked in close consultation with irrigators, the irrigation districts, the Irrigation Council, the Alberta Water Resources Commission, and the irrigation belt MLAs, developing computer simulation models of irrigation expansion scenarios and discussing their implications (Irrigation Water Management Study Committee 2002, 21). About a month before the process was concluded, the irrigation districts were given advance notice of what their expansion limits would be, providing them with the information they needed to apply for new water licences that would allow them to expand to these limits. In contrast, the Greens were not invited into the consultations and discovered it only very late in the process, during the public hearings for the federal environmental assessment of the Oldman Dam (Glenn 1999, 138–40). Thus, in the SSRB Water Management Policy, as in the Oldman Dam, the Greens were largely excluded from the policy process and could do little more than protest from the outside, after the most important policy decisions had already been made.

This exclusion is reflected in the design of the South Saskatchewan Basin Water Allocation Regulation, promulgated by Cabinet in September 1991 through an Order in Council. The regulation held that all unlicensed water in the SSRB was reserved to the Crown to be allocated according to the list of preferred water uses outlined in the Water Resources Act. Since irrigation was well-placed on that list, it was also well-placed to receive the unallocated water that would soon be made available with the completion of the Oldman Dam. Noticeably absent from the list were environmental water uses, which were not listed in the Water Resources Act and therefore not considered as valid candidates for receiving reserved water. In short, the basic effect of the reservation was to set aside unallocated water, by default, for domestic and economic uses.

The regulation also established expansion limits, expressed in terms of irrigable acreage, for each of the thirteen irrigation districts and for a number of private irrigation areas. Based on these expansion limits, it further outlined the maximum volume of water that each irrigation district and private irrigation area could claim, through new water licences, to support future expansion.[10] Since most of these expansion limits were higher – and some significantly higher – than the status quo, this greenlighted many irrigation districts to apply for new water licences and proceed with new expansions (Alberta Environment, "South Saskatchewan River Basin Water Allocation" 2003, 9–10). And since the districts had received advance notice of their expansion limits, they got a jump on applying for licences, before opposition could develop.

Finally, the regulation outlined minimum streamflows for a number of rivers in the Oldman sub-basin, but the levels were so low that many observers described them as "hardship levels," just enough to keep fish alive for short periods of time. The levels were also virtually unenforceable, as all water licences, including the newly issued ones, would have priority over instream flows during water shortages. Few Greens were impressed by the minimum streamflows, or with the regulation altogether. As one representative of Trout Unlimited summarized, "This [regulation] is in keeping with the way Alberta has been since the turn of the century, which is to say, 'You want water: we'll supply it'" (Barnett 1992).

By the end of 1992, the Aggies had secured both the Oldman Dam and the additional water that came with it. In May, the Alberta government rejected the report of the Oldman Dam Environmental Assessment Panel and it became clear shortly afterward that the federal government had little intention of trying to enforce the panel's recommendations. By this time, the dam was almost fully operational, and in August the Alberta government issued itself a final licence for the dam, signalling that it considered the dam to be a closed issue. The South Saskatchewan Basin Water Allocation Regulation was issued in September, and, by December, seven irrigation districts had received new water licences based on it, effectively allocating the new water made available by the Oldman Dam (Glenn 1999, 140). Friends of the Oldman River tried unsuccessfully to challenge the new water licences in court and continued to try to force Ottawa to implement the panel's recommendations,

10 It is important to note that irrigation districts are allowed to expand beyond their acreage limits but not their volumetric limits. That is, if a district can find water savings through efficiency and conservation measures, it can use this water to irrigate lands beyond the limits specified in the regulation.

but to little avail. Both the dam and the water licences had been secured through the Aggies' hard power in the water policy subsystem.

Summary

Though the election of a PC government in 1971 was an external shock to the water policy subsystem, this shock did not result in a move away from the hard water policy that Alberta had practised for decades under Social Credit governments. At the time of the shock, there was no opposing advocacy coalition to exploit it and work towards a policy shift. Moreover, the PCs moved quickly to align themselves with the irrigators and become the new political actors in the Aggie coalition. They embraced the hard approach to water governance, which aligned well with their more interventionist view of government in general, and adopted hard policy objectives that went well beyond that of their predecessors. This was clearly outlined in the 1975 Water Management for Irrigation Use policy, which, among other things, committed the government to constructing a dam on the Oldman River and using this new water to support future irrigation expansion. For the next seventeen years, the Alberta government worked to implement these commitments, encountering increasing political resistance along the way.

Throughout this period, Aggie hard power was decisive in advancing and maintaining the hard policy agenda. Irrigators, irrigation belt municipalities, irrigation-related industries, irrigation belt MLAs, and irrigation bureaucrats in the Agriculture and Environment Departments, shared common beliefs about the value of irrigation and the importance of irrigation expansion to the future prosperity of southern Alberta. Through organizations such as the Alberta Irrigation Projects Association, the Irrigation Council, and the irrigation caucus of the PC party, these actors worked in close concert with each other, sharing information, plotting strategy, securing resources, and advancing the cause of irrigation expansion. The Aggies' influence with policy decision-makers had to do with the importance of irrigation to the regional economy and, more significantly, the importance of the irrigation belt in the PCs' electoral base. The irrigation belt became even more important to the PCs through the 1980s as the Getty and Klein governments lost support in Edmonton and had to rely on the southern seats to maintain their majority government. Irrigation belt MLAs also served in key Cabinet and caucus positions, giving the Aggies a strong voice within the government's central decision-making bodies, ensuring that it would stick with the hard policy program, even after substantial political resistance materialized.

Political resistance came from a number of quarters during this period – landowners affected by dam projects, First Nations, and Greens – but only the last solidified into a concerted advocacy coalition. Many landowners were bought off by government compensation or absorbed into the Green movement, while First Nations periodically aligned with the Greens but preferred to deal with the province on a government-to-government basis. The formation of Friends of the Oldman River in 1987 marked the entrance of the Green advocacy coalition into the water policy subsystem, but this did not by any means open the doors to the water policy process. Instead, the Greens were treated as policymaking outsiders and, in their efforts to break into the policy process and influence policy decision-makers, they resorted to many of the tactics used by outsiders everywhere: public protests, public and media advocacy campaigns, and court action. Though they were ultimately unsuccessful in preventing the construction of the Oldman Dam and the allocation of more water to support irrigation expansion, their campaign against these hard policy measures would have a significant long-term impact on the policy subsystem, which we will explore in chapter 7.

Chapter Seven

The Water Act

Everything with rivers here is pre- and post-Oldman Dam.

Stewart Rood, biologist, University of Lethbridge

By the second half of 1991, the Alberta government was near the end of one water policy process and at the beginning of another. While the two were inter-related, they were also a study in contrasts.

The process nearing its end was the construction of the Oldman Dam. As recounted in chapter 6, it had originated in the summer of 1974 and the decision to build a dam was announced during the 1975 election campaign. No public consultations were held prior to the Oldman Dam decision, which was motivated, in large part, by the PCs' desire to lock up electoral support in the irrigation belt. Having committed itself to building a dam, and being subject to considerable political pressure from the Aggies, the Alberta government was never likely to back away from this decision. Various rounds of study and public consultation were undertaken in the years following the decision to build a dam, but these offered little opportunity other than protest to influence government decision-making. In this way, the Greens found themselves on the outside looking in at the Oldman Dam policy process.

The policy process that was just beginning was an overhaul of the province's water legislation. From the outset, it was qualitatively different from the Oldman Dam process. The Alberta government went out of its way to involve water policy stakeholders and the interested public at an early stage in the policy process. Public workshops were held, stakeholder advisory committees were formed, written and oral submissions were solicited, and even toll-free phone lines were set up to gather water policy input that would be used to inform the design, and later the revision, of new water legislation. Rather than commenting on policy decisions that had already been made, stakeholders were given the opportunity to participate in policy formulation and to shape water policy design. Moreover, all water policy stakeholders were welcomed into the policy process, transforming the Greens from political outsiders to political insiders who could now do much more than simply protest.

The change in approach was due largely to the government's experience with the Oldman Dam and its desire to contain future conflicts in the water policy subsystem. This change had huge implications for how water policies were made and what policy designs would result. Through the public and stakeholder consultations, the Greens now had more opportunities to use their soft power to push their soft water policy preferences. The Aggies found themselves on the defensive in the consultation processes but were able to use their hard power at the policy decision-making stage, once the proposed reforms entered the legislative arena. In this way, the Aggies successfully defended the water licences and water infrastructure in which they were heavily invested, while the Greens added a number of new soft policy instruments to Alberta's water legislation framework. The end result was a shift to soft*er* water policy, with soft reforms added to the hard status quo.

In this chapter, we examine the shift to softer water policy through the passage of the Water Act in 1996. We begin by identifying two shocks to the water policy subsystem in the early 1990s, one internal to the subsystem (the rise of the Greens and their resistance to the Oldman Dam) and one external to the subsystem (the Klein Revolution). Together, these shocks opened a path for major policy change, as predicted by the first ACF hypothesis on policy change. This path began with a review of Alberta's existing water legislation, proceeded through three drafts of new water legislation, and culminated in the passage of the Water Act. Throughout this process, the Greens used their soft power and the Aggies used their hard power to influence policy decision-makers, but the Greens enjoyed much more success than in the past because the new coalition opportunity structures afforded a more open and inclusive policy development process used by the Alberta government.

Internal and External Shocks to the Policy Subsystem

The early 1990s was a turbulent time in Alberta politics. The Oldman Dam conflict was at its height, with the Greens putting up unprecedented resistance to the Aggies' hard policy agenda, transforming the water policy subsystem from a staid one-coalition-dominant structure to a more conflictive two-coalition-pluralistic structure. Alberta's wider political system was also going through a major shake-up with the selection of Ralph Klein as the PC Party's new leader and the province's new premier in December 1992. This ushered in the "Klein Revolution," a radical neoconservative program designed to reduce the size and cost of government in order to deal with the province's deficit and debt problems. Considered together, these developments constituted two major shocks to the Alberta water policy subsystem, the emergence of the Greens constituting an internal shock and the Klein Revolution constituting an external shock. The shocks were mutually reinforcing in the sense that they benefited the Greens more than the Aggies. Moreover, the shocks opened up a path for major policy change, a path that eventually ended, through various twists and turns, with the passage of the Water Act in 1996.

The emergence of the Greens was an internal shock to the water policy subsystem because it reconfigured the basic political dynamics of water policymaking. Before the emergence of the Greens, policymakers could pander to the Aggies without fear of provoking a political backlash from rival advocacy coalitions. After the emergence of the Greens, however, politicians came to realize that siding with one coalition came with a political price. Conservative policymakers had learned this lesson the hard way during the Oldman Dam conflict. Their siding with the Aggies had provoked a massive Green backlash that had damaged their image on environmental issues and jeopardized their support with environmentally conscious voters. With two advocacy coalitions to contend with, politicians now had incentives to be more coalition-neutral – or at least appear more coalition-neutral – so as not to alienate either coalition. This new political reality was not immediately apparent when the Greens first emerged in the late 1980s, but became quite apparent during the Oldman Dam experience. So the shocking impact of the Greens' emergence took some time to materialize but, as it did, it changed the government's approach to water policy development in important ways.

One way in which policy development changed was an increased reliance on public and stakeholder consultations. Public and stakeholder consultations had been part of water policy development in Alberta since at least the 1960s, but became more frequent and impactful in the post–Oldman Dam era. The development of the Water Act, for instance, featured two rounds of extensive public and stakeholder consultations in four years. More importantly, these consultations were held in the early stages of the policy process before substantive policy decisions had been made, so that their impact can be seen in the final design of the legislation. Relying on public and stakeholder consultations allowed the Alberta government to maintain (the appearance of) coalition-neutrality, avoiding the alienation of either coalition.

Moreover, the Alberta government made a point of bringing Greens into stakeholder consultations by giving them designated seats in these processes. This was a major change to policy development, transforming the Greens from political outsiders, who could do little more than protest policy decisions after the fact, to political insiders with a real voice in the early stages of policy development. It also conferred a substantial degree of legitimacy on the Greens, who could now sit across the table from the Aggies as stakeholders of relatively equal status. In addition, the deliberative nature of stakeholder consultations played to the Greens' strengths. They could use their soft power to make persuasive arguments in favour of soft policy reforms, which, as we will see below, were compelling enough that many of these reforms were included in the reports and recommendations produced from the various stakeholder consultation processes.

In ACF terms, the government's increased reliance on public and stakeholder consultation processes shook up the water policy subsystem by realigning the coalition opportunity structures. The Greens, in particular, now had more opportunities to influence policy decision-makers than in the past. They still could not match the Aggies' hard power, but they now had more opportunities to employ their soft power, which they used to good effect in the design of the Water Act.

A further effect of the Greens' rise was the gradual emergence of Alberta Environment as a policy broker in the water policy subsystem. The Greens' resistance to the Oldman Dam created far more conflict than the Alberta government had ever expected or wanted, and one of its goals in the aftermath of this conflict was simply to "keep the peace" in the water policy subsystem. This was particularly the case after Ralph Klein, who had served on the front lines of the Oldman Dam conflict, became premier in late 1992. Since water policy was primarily within the ambit of the Environment Department, and the department already had a mixture of dam-building Aggies and river-protecting Greens, it became the government organization charged with keeping the peace. It oversaw the public and stakeholder consultations, ensured that both advocacy coalitions had seats at the table, and tried to find common ground between them. The department's brokerage role was particularly important in fashioning the SSRB Water Management Plan (as we will see later in chapter 8), but it was also formative in the development of the Water Act, which was fundamentally a compromise between the Aggies' and Greens' respective policy preferences.

Finally, the emergence of the Greens and their resistance to the Oldman Dam shook up the water policy subsystem by seriously challenging the credibility of the hard approach to water governance. For the Greens, the Oldman Dam was symbolic of everything that was wrong with water governance in southern Alberta: it destroyed riverine habitats, disrupted natural river flows, and facilitated the diversion of more water from a river that was already heavily allocated. Their environmental critiques resonated with a substantial portion of the public and found support from various expert sources, including the Oldman River Dam Environmental Assessment Panel. Even within the Alberta government, the continued credibility of the hard policy approach came into question as early as 1990, prompting a review of the province's water legislation and initiating the process that would lead to the eventual adoption of new legislation in the form of the Water Act.

The external shock, affecting the policy subsystem at roughly the same time, came in the form of the Klein Revolution. This was an external shock because it constituted a major shift in the governing party's governance philosophy, one that had an impact on all policy subsystems, including the water subsystem. By the early 1990s, the Alberta government had a considerable deficit and debt problem, due partly to a sluggish economy and partly to the Getty government's inability to reign in questionable spending. Upon assuming the premiership in December 1992, Ralph Klein opened the government's books to public scrutiny to show the extent of the fiscal problems and formulated a strategy for attacking them. This involved a hard turn to the ideological right and an embrace of the neoconservative principles of lower taxes, less spending, and smaller government. This approach to governance was very different from the selective interventionism espoused by Peter Lougheed and continued under Don Getty. Klein campaigned on a neoconservative platform in the June 1993 election and won a majority mandate to implement this program.

Between 1993 and 1995, the Klein administration oversaw a government restructuring that was unprecedented. "To refer to the government's attack on the deficit as 'aggressive' is something of an understatement"

(Barrie 2004, 265). Public sector wages, including those of the premier and Cabinet, were cut by 5 per cent, 1,000 public service positions were eliminated through lay-offs and early retirements, and a number of government functions were privatized, including the sale of alcohol. Just about all remaining parts of the government experienced deep funding cuts, including a $280 million cut to health care, a $245 million cut to education, a $140 million cut to post-secondary education, and a $100 million cut to social services (ibid.). Alberta Environment had its budget cut by 26 per cent and had to scale back many of its services (Parsons 1994). These painful cuts were vociferously resisted by public sector unions and social activists, but Klein stuck to the program, and the combination of expenditure reductions and increased revenues from a resurgent economy eliminated the deficit by 1995.

Klein's personal commitment to neoconservatism is questionable: he showed few signs of this ideological orientation as mayor of Calgary in the 1980s; he was regarded with suspicion as being too liberal by some PC caucus members when he was first elected as an MLA in 1989; and, once the deficit was eliminated, government spending rose again by 60 per cent between 1997 and 2001 (Martin 2002; Barrie 2004, 266). However, Klein was convinced that the deficit had to be defeated, he had campaigned on this promise, and this sustained his resolve to stay the course on the government cuts. He was also supported in the PC caucus by a group of committed, young neoconservative MLAs, first elected in 1993, who referred to themselves as the "deep six": Murray Smith, Lorne Taylor, Ed Stelmach, Jon Havelock, Lyle Oberg, and Mark Hlady. All but Hlady would go on to occupy senior Cabinet positions in Klein governments, ensuring that neoconservatism would continue to influence government policymaking even after the Klein Revolution had initially subsided (Martin 2002, 142).

In the water policy subsystem, the main effects of the Klein Revolution were to enhance the credibility of soft policy options and impair the credibility of hard policy options.

Some of the most prominent soft policy instruments are those that put a market value on water, through either full-cost pricing or water licence trading. These types of policy instruments had been discussed and debated within the water policy subsystem for many years, but had never gained much traction. Economists lauded them, but Greens were suspicious of introducing market forces in water governance, and Aggies did not want to shoulder the additional production costs that these instruments might entail. For neoconservatives, however, market-based policy instruments are inherently superior to state ownership and regulation. So, in the context of the Klein Revolution, full-cost water pricing and water licence trading came to be seen as credible and viable policy options. As shown in the content analysis in chapter 4, these policy options went from being barely mentioned in the 1978 stakeholder consultations to being mentioned by more than three-quarters of the participants in the 1995 stakeholder consultations. Moreover, water licence trading (but not full-cost water pricing) came to be supported by the Aggies in the early 1990s, considerably boosting its chances of being included in Alberta's water policy reforms.

Hard policy options, in contrast, were not very compatible with neoconservatism, and neoconservatives attacked them as being financially unviable. The water management infrastructure that was the centrepiece of Alberta's hard policy was very expensive to construct and maintain, and the vast majority of these costs were borne by the provincial government. The Oldman Dam alone cost the Alberta government $353 million in 1986 dollars ($900 million in 2012 dollars) and, by 1993, the province was spending an additional $19 million every year in its infrastructure refurbishment program (Glen 2012; Parsons 1994). For a government battling a deficit, these costs were not insignificant and the Aggies did not escape the Klein Revolution cuts. The refurbishment program was cut by about $2 million per year and, because it was a cost-sharing program, irrigators' share of refurbishment costs was increased from 14 to 25 per cent (Parsons 1994). The neoconservative emphases on smaller government and less government intervention also undermined any plans to expand the water management infrastructure further. In this way, neoconservatism impaired the credibility of hard policy on financial grounds just as the Greens were impairing it on environmental grounds.

All of these developments affected the subsystem enough that a path for major policy change opened up. This path ended in the passage of the Water Act, the first major reform of Alberta's water legislation in nearly a century and the turning point in Alberta's shift to softer water policy. In the remainder of this chapter we examine the development of the Water Act in three phases, each defined by a different piece of draft legislation. In all phases, the advocacy efforts of the coalitions are recounted, and the picture that emerges is one of Green soft power versus Aggie hard power, similar to what we have seen in previous chapters. This time, however, with their soft power growing, with more favourable coalition opportunity structures, with a policy broker keen to placate them, and with a somewhat discredited policy status quo, the Greens were able to use their soft power to greater effect and to have some soft policy instruments included in the new legislation. Yet the Aggies were still a formidable political force that held the balance of hard power in the policy subsystem, and they used this power to preserve the hard policy status quo that they had helped build over many decades. The end result, in the Water Act, is the addition of new soft policy instruments to the preserved hard policy legacies in a sort of policy sedimentation that can be best characterized as a shift to soft*er* water policy.

The Water Management and Conservation Act

In 1990, a group of senior officials from the Departments of Environment, Agriculture, Municipal Affairs, and Transportation and Public Works, acting through the Alberta Water Resources Commission, advised the government that the Water Resources Act was in need of review. An official familiar with the recommendation explained that this was largely in reaction to the Oldman Dam controversy and the fact that further dam construction, in support of irrigation expansion, had become too politically difficult. Other means were needed to manage water scarcity in southern Alberta, and the Water Resources Act, which was designed mostly to encourage water development, provided few management options. Months later, in March 1991, Environment Minister Klein publicly announced the government's intention to review and revise the Water Resources Act, aiming to have new water legislation ready for early 1992 (Glenn 1999, 108). It is not clear, at this point, whether the government intended a major overhaul or a fine-tuning of the legislation, but the door to water policy change had been opened.

The first step in this process was the release of a public discussion paper in July of 1991 entitled "Water Management in Alberta: Challenges for the Future." Produced by Alberta Environment, the paper made the case, to Albertans generally and to water policy stakeholders specifically, for pursuing water policy reform: "Water management has changed and new policies and programs have been developed to promote the wise use of this precious resource. However, much remains to be done. The Water Resources Act is essentially the same as it was in 1931. It is still primarily a mechanism for allocating water until supplies are exhausted. The Act must be changed to reflect changing water management circumstances in Alberta and to prepare for future challenges."

The paper also hinted at the direction legislative reform should take. Touting the need to use the province's water resources wisely, it stated that "wise use implies water conservation and demand management and recognizes that sustainable water management practices must be pursued" (Alberta Environment 1991, 15). Nothing in the discussion paper implied that the existing hard water infrastructure would be dismantled or that existing water licences would be cancelled, but there was a clear call for a shift to softer water policy.

The discussion paper also outlined a two-stage process for legislative reform, both involving extensive public and stakeholder consultations. The first stage would collect public and stakeholder input for consideration in the drafting of new water legislation, while the second stage would involve a public and stakeholder review of specific draft legislation (Alberta Environment 1991, 23–4). The Alberta government stuck to this process, though the initial timelines proved wildly optimistic; it took over three years to complete rather than the one year initially envisioned. Managing the process throughout was Alberta Environment, organizing meetings, providing background papers, answering questions, collecting input, and synthesizing this input for policymakers.

The first stage was initiated with the release of the discussion paper and involved a wide range of consultation mechanisms: the discussion paper itself had an attached response sheet that could be mailed to Alberta Environment free of charge; a toll-free telephone line was established to receive input and field questions; "awareness meetings" were set up with stakeholders around the province to provide instruction on how they could participate in the reform process; and, most importantly, public workshops were held at fourteen locations throughout the province so that stakeholders and interested members of the public could submit oral and written statements on their preferred policy options (Alberta Environment 1991, 23–4). These consultations were far more extensive than anything previously seen in the water policy subsystem, and they occurred at an earlier stage in the policy process, collecting input for policy formulation rather than reviewing policy decisions that had already been made.

As expected, the first stage of consultations in late 1991 and early 1992 was characterized by a strong "Aggies versus Greens" dynamic. At the public workshops, representatives of both coalitions were omnipresent, the Greens making a case for a major legislative overhaul and the Aggies arguing for the maintenance of most of the policy status quo. Both coalitions were also active outside of the workshops, publicly staking their policy positions and attempting to show the breadth and depth of support behind them. For instance, on 9 January 1992, Unifarm, the province's largest farmer advocacy organization, passed resolutions at its annual meetings calling for the protection of irrigators' existing water rights in any legislative reforms. The very same day, a coalition of thirty-three environmental groups released a joint position paper calling for increased minimum streamflows and the cutting of water licences to achieve them, if necessary (Marshall 1992). This kind of Aggie-Green tit-for-tat became a common feature of the legislative reform process.

The public workshops, in particular, played to the Greens' strengths and became important vehicles for transmitting their policy preferences to Alberta's water policymakers. The workshops provided open forums in which many different water policy ideas could be discussed and their merits openly debated. In this deliberative atmosphere, the Greens marshalled considerable evidence for the decline of riverine environments and made very persuasive arguments for arresting these declines. The Greens were also quite successful at mobilizing their supporters to attend the public workshops, giving them a presence that made them seem very much the Aggies' equal. The Aggies countered the Greens' arguments by pointing to the communities and jobs relying on the appropriation of southern Alberta's water, arguments that carried considerable weight, but were also tinged with material self-interest. Given how far removed the workshops were from electoral politics, though, the Aggies were unable to dominate them with their hard power, and they found themselves in the unfamiliar position of being on the defensive for much of the time.

The influence of the Greens and their soft power in the public workshops is best illustrated through the results of a questionnaire distributed to the workshop participants. The questionnaire results lend overwhelming support to the Greens' policy preferences. When asked whether new legislation should "reserve a minimum flow level in all rivers and lakes which cannot be allocated to users," 90 per cent of respondents agreed. Similarly, 89 per cent of respondents agreed that instream flow requirements should be established for all major rivers and streams, and 79 per cent supported the idea that instream flow volumes for purposes such as fish habitat protection and recreation should be protected through licensing. Even the results relating to market-based policy instruments closely mirrored the policy preferences of the Greens: only 32 per cent of respondents supported the introduction of water licence trading, while 85 per cent supported the introduction of the "user pay concept" (Alberta Environment, Environmental Council of Alberta, and Water Resources Commission 1992, Appendix B).

Though the workshops concluded at the end of 1991, it was not until August 1994 that new draft legislation was released by Alberta Environment. In early 1992, the input collected from the response sheets, the toll-free line, and the public workshops was synthesized in a joint report by Alberta Environment, the Environment Council of Alberta, and the Alberta Water Resources Commission. After that, the process seemed to stall. It is not entirely clear why it took the department more than two years to produce draft legislation, but the effects of the Klein Revolution were undoubtedly part of it. The department's budget was slashed, personnel were laid off, and both the Environment Council and the Water Resources Commission were discontinued and folded into the department. Another part of the explanation likely lies in the complexity of the task itself. Rewriting the province's water legislation was an immense and complicated task, as evidenced by the draft legislation itself, which constituted an eleven-part, 185-page booklet when it was finally released. Nevertheless, some stakeholders, particularly Greens, grew frustrated with the delay: "Input provided by the public in 1991 disappeared into the bowels of the civil service in Edmonton and nothing but speculation about what was going on was heard until release of the draft legislation in August 1994. The Department has failed to follow its own 'Public Involvement Guidelines'" (C.E. Bradley to Alberta Environmental Protection 1 March 1995).

Though it never became law, the draft Water Conservation and Management Act established a general template for water policy reform that proved enduring: existing water licences and hard infrastructure were preserved while adding a variety of soft policy instruments to provide new tools for managing water scarcity. This basic compromise between Aggie and Green policy preferences would remain prevalent throughout the reform process.

The discussion draft made it very clear that the existing water licences of irrigators and other water users would be protected under the new legislation. Existing licences were not only recognized, but their original terms were defined as superseding the new act, thereby limiting its impact on irrigators' water use practices. As the notes accompanying the draft legislation explained, "This provision confirms the Government's commitment that existing rights which are in good standing will be protected. This commitment is in keeping with the long history of water rights and the significant investments which have been made in relation to those water rights" (Alberta Environmental Protection 1994, 35).

The draft legislation also preserved the "first in time, first in right" approach to prioritizing water licences, leaving the very large and senior licences of the irrigation districts intact and ensuring that any future water licences would be junior to these established uses (Alberta Environmental Protection 1994, 38).

The draft legislation also included some unprecedented soft policy instruments. The director of the new act – a senior official within Alberta Environment – was empowered to impose moratoria on new licences in over-appropriated areas, and the Crown was empowered to reserve unallocated water "for a variety of purposes including instream protection" (Alberta Environmental Protection 1994, 64 and 79). The moratoria and reservation instruments allowed government regulators to halt (or at least slow) continued growth in water diversions, establishing regulatory boundaries between human and environmental water uses. The discussion draft also included provisions to allow market trading of water licences under strict conditions, a major change to the way water was reallocated in the province. As part of this, the Act empowered the government to withhold up to 10 per cent of the volume of any water licences traded and to reallocate this water for environmental protection purposes (Alberta Environmental Protection 1994, 99). These "conservation holdbacks" were the first policy instrument proposed to allow for the reallocation of water from economic to environmental uses and, although the volumes of reallocated water were anticipated to be small, their inclusion was a policy victory for the Greens.

In short, the Water Management and Conservation Act established an Aggie-Green compromise in which the hard policy status quo would be largely maintained but overlain with a number of new soft policy instruments. Neither coalition was entirely satisfied with this policy compromise, and the rest of the legislative reform process involved efforts by the Aggies to minimize the soft policy changes and the Greens to eat away at the hard policy status quo. The only exception to this pattern was in water licence trading, a soft policy instrument that was accepted by most Aggies but not by most Greens. The Aggies accepted it – some of them reluctantly – because there was considerable enthusiasm for introducing market-based water reforms among the neoconservatives in the Klein government, some of whom were also Aggies. The irrigators calculated that if they had to embrace a market-based instrument, licence trading was preferable to full-cost water pricing, as the former gave irrigators a valuable asset while the latter imposed additional costs on them. The conservation holdback provision made water licence trading somewhat more palatable for the Greens, but many still regarded the introduction of market forces in water governance with considerable suspicion and opposed licence trading to the very end.

The Water Act, Part 1: Kill Bill 51

The public release of the draft Water Conservation and Management Act in August 1994 initiated the second stage of consultations envisioned in 1991, the public feedback stage. To guide the consultations, a Water Management Review Committee was established and given the tasks of reviewing the draft legislation, receiving public input, and making recommendations to Cabinet. The committee was chaired by PC MLA Glen Clegg, who represented the northern riding of Dunvegan and, as such, was a neutral between the Aggies and Greens. The rest of the committee represented a wide range of stakeholders in the Alberta water policy subsystem, including environmentalists, conservationists, recreationalists, tourism operators, industrialists, irrigators, utility operators, academics, lawyers, and municipal representatives (Water Management Review Committee 1995, 97–8). The committee had strong representation from both advocacy coalitions, and its work reflects this balance.

Similar to its approach during the first stage of consultations, the government went to considerable lengths to collect public and stakeholder input on the draft legislation. The public consultations involved "public review sessions in 14 locations in Alberta, a toll-free information line, written submissions and a two-day workshop for the many interest groups who had been involved in the 1991 consultation process" (Water Management Review Committee 1995, iii). Over 1,500 people participated in the public review sessions, 3,500 people attended the information sessions, and over 300 written submissions were received. (These submissions were sampled and analysed as part of the content analysis in chapter 4.) The committee members also deliberated at length over the discussion draft and tried to reach consensus recommendations wherever possible. Where consensus proved impossible, "strongly supported" recommendations were made, with dissenting views noted in the final report (Water Management Review Committee 1995, iii).

The Greens benefited from this process because of its open, deliberative nature, largely insulated from electoral and legislative politics. Green groups, such as the Alberta Fish and Game Association, the Southern Alberta Environment Group, Friends of the Oldman River, Ducks Unlimited, Trout Unlimited, and the Environmental Law Centre had representation on the committee. These and other Greens also made submissions to the committee and participated in the two-day workshop for interest groups, ensuring that Green views were strongly represented in the committee's deliberations (Water Management Review Committee 1995, 97–105). Public input also showed strong support for Green policy positions, a fact remarked upon consistently throughout the committee's final report (ibid.). The Aggies also had strong representation on the committee, with members from the Alberta Irrigation Projects Association, Unifarm, and the Alberta Cattle Commission, and there were many public submissions in support of their policy positions (98). But the process was so far removed from legislative politics that they could not use their hard power to dominate the proceedings. Instead, the committee proved to be a soft-power-friendly environment where public opinion, information, and persuasion were of most use. For the Aggies, this constituted a major decrease in policy influence, while for the Greens it was a major increase.

The *Report of the Water Management Review Committee*, submitted to Environment Minister Lund in July 1995, reflects this substantial Green influence, carrying forward the soft policy instruments proposed in the discussion draft and pushing them even further. In addition to the moratorium, Crown reservation, licence trading, and conservation holdback instruments proposed in the discussion draft, the Water Management Review Committee also recommended that outright licence expropriation (with compensation) be included in the new legislation: "The WMRC recommends the new water legislation enable the Government to take over a water licence for specific purposes, including implementing minimum aquatic and riparian ecosystem protection requirements. When a licence is taken over there should be a mandatory entitlement to compensation [for the licence holder]" (Water Management Review Committee 1995, 20).

It should be noted that this was a "strongly supported" but not unanimous recommendation from the committee, with some Aggies dissenting on the grounds that licences should not be expropriated under any circumstances, and some Greens dissenting on the grounds that expropriation should be permitted for all environmental restoration efforts, not just for "minimum" requirements (Water Management

Review Committee 1995, 20). However, it is still remarkable that this recommendation was included in the report because, for the first time, serious consideration was given to how significant amounts of water might be reallocated from economic to environmental protection purposes. In this way, the Water Management Review Committee recommendations were practically revolutionary and constituted a high-water mark for Green influence in the legislative reform process.

In light of the Water Management Review Committee report, officials in Alberta Environment revised the draft legislation, changed its title, and forwarded it to Cabinet. The title change was based on a recommendation of the Water Management Review Committee who argued that "the new water legislation deals with a broader range of issues than just water conservation and management" (Water Management Review Committee 1995, 4). A few weeks later, in November 1995, Bill 51, the new Water Act, was given first reading in the Alberta legislature.

Almost immediately, the irrigation sector was in uproar over its contents. Bill 51 provided the environment minister with considerable discretionary authority, including the authority to establish instream flow needs to protect riverine environments, and to prioritize these instream flow needs ahead of all consumptive water licences, regardless of their seniority. In accordance with the Water Management Review Committee recommendations, Bill 51 also gave the minister the power to expropriate water licences (with compensation) to find water to meet instream flow needs on rivers that were already heavily allocated. For irrigators, the expropriation power and other provisions in Bill 51 left their water licences vulnerable to the whims of government regulators who, bowing to Green pressure, could take their water at any time and reallocate it to the environment. This was unacceptable and they immediately mobilized to block the bill's passage.

Now that the policy process had entered the legislative arena, the Aggies were finally able to bring their hard power to bear on the proposed water reforms. During the late fall of 1995 and early winter of 1996, irrigators, ranchers, and other agricultural interests rallied under the umbrella of the Alberta Cattle Commission and persuaded irrigation belt MLAs, the minister of agriculture, and the Department of Agriculture to join in their lobbying efforts. The primary target of their lobbying was the minister of environmental protection, Ty Lund, who represented the rural riding of Rocky Mountain House and was himself a third-generation Alberta farmer (Legislative Assembly of Alberta, 8 May 1996, 1718). Lund was not an ideologically committed Green and proved vulnerable to the Aggies' lobbying efforts. The Greens tried their best to rally public support for Bill 51 but simply could not match the legislative influence of the Aggies, nor counteract the ideological congruence of the Aggies and key government decision-makers. In the end, Bill 51 never made it past first reading in the legislature and was allowed to die on the order paper (ibid., 1717). The Greens had scored some success in having Bill 51 introduced, but Aggie hard power had reasserted itself in the legislative arena after having been muted in the public consultation processes.

The Water Act, Part 2: Institutional Layering

Despite Bill 51's demise, the Alberta government remained committed to updating its water legislation and introduced a new water bill in April 1996. The new bill, known as Bill 41, eliminated the expropriation power, included strong protections for existing water licences, and was much more to the Aggies' liking. Bill 41 was endorsed by just about every agricultural organization in the province and was supported in the legislature by the irrigation belt MLAs, giving it the Aggies' imprimatur (Legislative Assembly of Alberta, 14 August 1996). The Greens, in contrast, were disappointed that the expropriation power had been dropped from Bill 41, leaving conservation holdbacks as the only instrument for reallocating water from existing licences to the environment, and they voiced their concerns publicly. The Opposition Liberals moved an amendment to restore the expropriation power, but the motion was soundly defeated by the government (Legislative Assembly of Alberta 27 August 1996, 2442–7). Ultimately, the Aggies' hard power carried the day, and Bill 41 passed the Alberta legislature with relative ease on 27 August 1996.

The new Water Act contained the same basic policy compromise that had been proposed in the draft Water Conservation and Management Act of 1994: the legacies of the policy hard era, in which the Aggies were heavily invested, were retained, but were overlain by a variety of new soft policy instruments, most of which were advocated by the Greens. In this way, the policy shift has been one of degree rather than of kind, because the water supply infrastructure and the prior allocation system inherited from the hard policy era have not been abandoned. This is what Streeck and Thelen (2005) call "institutional layering" as new policy rules are added on top of old policy rules in a sort of policy sedimentation (22–4). Though the change was sedimentary rather than sweeping, the cumulative change has been quite significant, particularly in a policy subsystem where the hard policy approach had prevailed for over a century.

The sedimentary nature of the Water Act reforms is particularly evident in its distinction and treatment of existing licences and new licences. Existing licences are those that existed at the time the legislation was passed, while new licences are those that have been issued since the legislation took effect. The Act states that existing licensees "can continue to divert water in accordance with their original priority, the terms and conditions of their original licence and the new Act. However, if there is a conflict between a term of a deemed licence and the new Act, the term of the licence prevails over the Act" (Percy 1996–7, 229). This, in effect, protects existing licensees – overwhelmingly irrigators – from some of the provisions of the Water Act that might have threatened their licences. New licensees, in contrast, are treated very differently. Any new water licences distributed under the Water Act are subject to fixed terms and can be denied on the grounds that they contravene an approved water management plan (discussed below) or would negatively affect a riverine environment. New licences are also subject to a public notification process and can be challenged at the Environmental Appeal Board (237–8). In short, existing licences were preserved, and new licences – subject to much stricter treatment – were layered on top.

When it came to new water licences, the Water Act went even further, allowing for the distribution of new licences to be frozen altogether. The new legislation gave its director the authority to impose moratoria on the issuance of new water licences in heavily allocated streams. It also empowered the environment minister to reserve unallocated water for any purpose, including the preservation of instream flows (Percy 1996–7). This was first proposed in the draft Water Management and Conservation Act and was a substantial change from the 1991 SSRB Water Allocation Regulation, which had reserved all unallocated water for human water uses only.

One of the greatest changes brought about by the Water Act was the introduction of water licence trading. It permitted both temporary and permanent transfers of water licences, subject to some conditions. It was envisioned that water trading would be introduced on a basin-by-basin basis and would be limited to basins where water was already scarce, such as the SSRB. Accordingly, the Act called for the creation of basin water management plans that, among other things, were to include specific rules for water trading, developed through public and stakeholder consultations and approved by Cabinet. It was through the basin water management plans that regular, routinized water trading would be introduced. In the absence of such plans, water trading was still permitted, but each trade was subject to Cabinet approval (Percy 1996–7, 236). With these reforms, Alberta became just the second jurisdiction in Canada – BC was the first – to allow water entitlement trading. As we will see in chapter 8, a (partial) basin water management plan was approved for the SSRB in 2002, allowing water trading to begin there in earnest.

In conjunction with water licence trading, the Water Act also introduced conservation holdbacks as a policy instrument for recovering water for environmental purposes. A conservation holdback is a bit like a water tax on a water trade: it is a small volume of water that is held back from a water trade and reallocated for environmental use. Specifically, the Water Act empowers its director to withhold up to 10 per cent of the volume of any water traded in order "to protect the aquatic environment or to implement a water conservation objective, if the holdback is authorized in an approved water management plan" (Percy 1996–7, 239). For the first time, Alberta's water law had a provision allowing for any amount of water – however small – to be taken from existing licences and returned to the environment. This policy instrument was also regarded as central to the achievement of the water conservation objectives that would be outlined in the basin water management plans.

Essentially, the water conservation objectives were designed as instream flow targets, to be formalized in basin water management plans, and intended to maintain or restore healthy riverine environments. While the Environment Department had been studying and pursuing various instream flow objectives since the 1970s, as shown in chapter 4, the water conservation objectives of the Water Act were different because they were substantively linked with water licences. All new water licences, for instance, were to be distributed on the condition that they not impair the achievement of a water conservation objective. In this way, the new water conservation objectives would have the potential to limit water-takings on streams that were not already fully allocated, such as the Red Deer River. Water conservation objectives would have no impact on existing licences, except when existing licences were traded. During water trades, they were to serve as important benchmarks in the director's decision-making on whether (or not) to take conservation holdbacks.

The Water Act was a substantial shift to softer water policy in southern Alberta, but its passage still left much to do. The Act created soft policy instruments allowing for licensing moratoria, licence trading, water conservation objectives, and conservation holdbacks, but their introduction still depended on the formulation of basin water management plans. Of all the basins in Alberta, the priority for the Alberta government was the formulation of a plan for the SSRB, where water scarcity was being felt most acutely. From 2000 to 2006, over two phases involving extensive stakeholder consultations and negotiations, a basin water management plan was negotiated for the SSRB, and it is to this process that we turn our attention in chapter 8.

Summary

In the early 1990s, the Alberta water policy subsystem was buffeted by twin shocks, one internal and one external, which opened a path for major policy reform. The internal shock was the emergence of the Greens and the conflict over the Oldman Dam, which challenged the credibility of the hard policy status quo and forced the Alberta government to embrace a more open and inclusive approach to water policy development. The external shock was the Klein Revolution, a sharp turn to the ideological right by the governing Conservatives, which enhanced the perceived viability of soft policy options and impaired the perceived viability of hard policy options. These shocks affected the policy subsystem just as the balance of soft power began to tilt in favour of the Greens, so that they were well-positioned to push for their favoured soft policy reforms. In short, the internal and external shocks played key roles in the shift to softer water policy, as predicted in the first ACF policy change hypothesis.

Though the Greens became more influential in water policy development, the Aggies still retained the preponderance of hard power in the policy subsystem and, as such, remained the dominant coalition. Once the policy process moved beyond public and stakeholder consultations and into the legislative arena, the Aggies were able to bring their hard power to bear. They ensured that existing water licences were fully protected and that any powers allowing regulators to expropriate water licences for environmental protection purposes was struck from the legislation. In this way, the Aggies preserved much of the hard policy status quo; neither their water licences nor the infrastructure supporting these licences was in any way dismantled by the new Water Act.

The result was the layering of new soft policy instruments on top of the hard policy status quo, creating an overall softer approach to water governance.

The development of the Water Act was the first policy process in which both the Aggies and Greens were given seats at the policy development table, but it was not the last. This feature would be even more prominent in the next phase of water policy development in the SSRB, the creation of an SSRB Water Management Plan, a process that we will examine in greater detail in chapter 8.

Chapter Eight

The SSRB Water Management Plan

The most significant outcome was the conclusion that the limit of the surface water resources for the Oldman, Bow and South Saskatchewan River sub-basins had been reached. This was an important decision in a basin with a booming economy and increasing population. It sent a powerful message that more care, innovation and creativity are required in the management of allocated water. Residents of the SSRB will have to learn to manage their needs and the supply within a fixed limit.

Alan Pentney and Doug Ohrn, Alberta Environment

In water policy development, a drought can change everything. In the summer of 2000, southern Alberta was beset by a serious drought, one of the worst in the past century. The drought persisted into 2001 and even into 2002 in some parts of the SSRB. The drought's onset came just over a year after the Water Act had taken effect, and just as the Alberta government was gearing up for the development of the SSRB Water Management Plan. Given that the new policy instruments created by the Water Act were intended to help water users in southern Alberta cope with water scarcity, the drought raised the stakes of the plan's negotiation and added considerable urgency to the plan's conclusion. It also offered a stiff test of the government's commitment to the new, softer approach to water governance. The drought sparked calls for the construction of new pipelines, new dams, and even a new inter-basin diversion to help cope with water scarcity in the south, and there was a real potential that the Alberta government might abandon its softer course and revert to more familiar hard policy options.

The fate of the SSRB Water Management Plan and any new hard infrastructure program would be determined in the Alberta water policy subsystem, which continued to be characterized by the Aggie-Green dynamic that had emerged over a decade previously. The main difference now was the generally accepted place of the Greens in policy development processes. They were no longer political outsiders excluded from policy consultation, but legitimate stakeholders brought into policy processes at an early stage. This gave them status that was relatively equal to the Aggies', but they were still clearly the minority coalition. While the Aggies continued to draw on the hard power of their electoral and legislative clout, the Greens still relied on soft power in their efforts to influence policy decision-makers. Thus, the SSRB water management planning process was characterized by the familiar contest between Aggie hard power and Green soft power, and the design of the plan can be explained largely in terms of Aggie-Green compromises.

Positioned between the coalitions, and charged with stewardship of the planning process, was Alberta Environment, whose brokerage role was central to the plan's conclusion. Even more than in the Water Act process, the department took on the role of policy broker, bringing the coalitions together on co-management committees, providing them with a common knowledge base, and helping them to find negotiated agreements on contentious issues. In this way, the brokerage actions of the department were at least as important as the actions of the coalitions in forging the compromises that would eventually constitute the SSRB Water Management Plan.

A Policy Path Opens

By 1999, a new path for water policy reform was opening in southern Alberta, and both the Aggie and Green coalitions were poised to take advantage of it. This path was based on past water policy commitments and, as such, constituted an internal perturbation of the water policy subsystem that offered new opportunity for policy change.

When the SSRB Water Allocation Regulation was introduced in 1991, it included a requirement that the regulation be reviewed in the year 2000. You may recall from chapter 6 that the regulation introduced a number of water allocation measures in southern Alberta: it reserved all unallocated water in the SSRB for economic uses, it established expansion limits for southern Alberta's thirteen irrigation districts, and it specified minimum streamflows for a number of rivers in the Oldman sub-basin. The object of the year 2000 review was to evaluate the 1991 expansion limit and minimum streamflow levels to determine whether they were effective and acceptable. Alberta Environment, who would be leading the review, began gearing up for it as early as 1998. They intended the review to be a wide-ranging exercise that would model natural and actual flows in the SSRB and determine instream flow needs on the major rivers. It was also envisioned that stakeholders would be deeply involved in the review process. Two basin advisory committees were to be formed, with representation from all major stakeholder groups, one for the heavily allocated Oldman sub-basin, and one for the remainder of the SSRB (Alberta Environment 2000).

Just as the preparations for the year 2000 review were ramping up, the new Water Act took effect on 1 January 1999 and swamped the review process. The Water Act had its own scheduled imperatives, including a requirement that a water management plan for the SSRB be completed within three years of the Act coming into force. The Alberta government now had to contend with two scheduled water policy processes, and its solution was to merge them so that they satisfied both the regulation and the Water Act requirements. In this way, the nascent year 2000 review quickly morphed into the SSRB Water Management Plan process and became even more wide-ranging and consequential (Alberta Environment 2000). It was now not merely a policy review, but a planning exercise focused on "high priority quantity issues in the basin concerning water allocation and river flows" (Pentney and Ohrn 2008, 382). In short, past policy commitments came to fruition and shook up the water policy subsystem, opening another path for policy change.

As described in chapter 7, Alberta Environment had emerged as a policy broker in the wake of the Oldman Dam conflict, assigned to keep the peace between the warring Aggie and Green coalitions. The department had performed this role with some success in the processes leading to the Water Act and was asked to reprise this role in the development of the SSRB Water Management Plan. In fact, the department's brokerage role became even more pronounced in the new planning process and was one of the key factors in its successful conclusion.

As a government-endorsed policy broker, Alberta Environment was largely responsible for the design and oversight of the water planning process. An interdepartmental Steering Committee was created to guide the process, with eleven representatives from Alberta Environment, Alberta Agriculture, Food and Rural Development, and Alberta

Sustainable Resource Development. The federal Department of Fisheries and Oceans also had a representative, but with observer status only (Alberta Environment 2006, 2). The Steering Committee was designed to ensure that all relevant parts of the Alberta government would buy in to the plan, but most of the heavy lifting in plan development was done by Alberta Environment. It did most of the relevant research and advising on the plan, and it was a handful of Alberta Environment officials who chaired the Steering Committee, who undertook most of the necessary coordination, who drafted the eventual plans, who organized the public consultations, and who shepherded the draft plans through to Cabinet.

As originally intended in the year 2000 review, Alberta Environment opted for a co-management style of process with extensive stakeholder involvement. Rather than the two basin advisory committees originally envisioned, the department decided to create four basin advisory committees, one for each of the sub-basins in the SSRB. Representatives from municipalities, First Nations, major water-using industries, recreation groups, fish and game clubs, environmentalist organizations, irrigation districts, and agricultural organizations were invited to participate on the basin advisory committees, with all but the First Nations accepting the invitations. The First Nations demanded to deal with the Alberta government on a government-to-government basis, a demand that was eventually acceded, allowing them some input in the planning process (Alberta Environment 2006, 28). Aggies and Greens were both strongly – and purposely – represented on the basin advisory committees, and the committees were designed to work on a consensus basis so that the competing coalitions would be forced to find common ground. Each basin advisory committee was charged with developing a draft plan for its sub-basin. The draft plans were advisory but would be weighted heavily by the Steering Committee, which was ultimately responsible for putting together the total basin plan. Then, the draft plan would be reviewed by the public, revised by the Steering Committee, and forwarded to Cabinet for final approval.

Much like the early development of the Water Act, the SSRB water planning process created coalition opportunity structures that played to the Greens' strengths. The Aggies and Greens had relatively equal representation on the basin advisory committees and were treated as equally legitimate and welcome participants in the planning process, a status that the Greens had only recently acquired. The consensus decision rule meant that Green views could not be ignored in committee deliberations and would very likely inform committee recommendations. Moreover, the basin advisory committees were far removed from electoral and legislative politics, so the Aggies had a difficult time using their hard power to dominate committee deliberations. The Greens, however, benefited from the deliberative nature of the committees and could rely heavily on information (from a series of background reports) and public opinion (from the public consultation processes) to buttress their arguments and draw people towards their policy positions. In this way, the Greens were able to exercise more influence in the SSRB water planning process than might be expected of a minority advocacy coalition with no hard power in the policy subsystem.

Just as the planning process was about to get underway in 2000, the water policy subsystem was hit by an external shock: the 2000–2 drought[11] (Rannie 2006, 263). The drought increased the urgency with which both irrigators and environmentalists pursued their water policy objectives, and it had an unexpected political impact, particularly on the environment minister of the time, Lorne Taylor.

Taylor represented the southeastern riding of Cypress–Medicine Hat, one of the driest ridings in the province, and one particularly hard-hit by the drought. Access to reliable water supplies was a perennial problem in many rural communities in Taylor's riding. To rectify this, in 2001, a number of them banded together to form the Southeast Alberta Water Co-op to purchase water and secure a water pipeline to the water-stressed communities (Teel 2001). Under the Water Act, water licence trading was now permitted, but until an SSRB water management plan was finalized that would introduce standard rules for water

---

11 As the drought entered its second year, Alberta Environment forecasts predicted water shortages for the 2001 irrigation season in the upper Oldman sub-basin, meaning there would not be enough water available to meet all existing licences. It was estimated that at least forty-five junior licences were in jeopardy of being cut off from water, including a number of municipalities, large ranches, and an economically important food-processing plant near Taber. For irrigators who held the senior licences in the area, cutting off these junior licensees was extremely problematic because they lived in these municipalities and relied on some of these ranchers and food processors to buy their crops. Under the encouragement of Alberta Environment officials, most of the irrigation districts and other senior licence holders in the area came together and formulated a voluntary water sharing plan, approved by Alberta Environment, in which they agreed to "share the shortage" with junior licensees so that none would be completely cut off from water. In the end, all participants received about 60 per cent of their licensed entitlements (Rood and Vandersteen 2010, 1615–16).

trading in the basin, all trades required Cabinet approval. With this and other water trading schemes in mind, Taylor instructed his officials to make the introduction of water trading a priority in the SSRB water management planning process, prompting them to split the planning process into two phases: the first phase dedicated to water trading, to be undertaken almost immediately, and the second phase dedicated to water conservation objectives, to be undertaken in about a year's time. By October 2001, the first phase was underway.

The drought also opened a path for other water policy measures, particularly some hard policy projects long-favoured by the Aggies. In addition to the pipeline proposed by the Southeast Alberta Water Co-op, three other ambitious hard projects found their way onto the policy agenda in 2001, at the height of the drought: an irrigation expansion plan, a new dam proposal, and a renewed study of inter-basin diversions. The main advocate for these proposals was Environment Minister Taylor, who, despite his portfolio, had been a long-time enthusiast of hard water projects. Except for the Southeast Albert Water Co-op's pipeline, none of these projects got past the proposal stage, illustrating the extent to which the Alberta government had turned away from hard policy solutions in its efforts to manage water scarcity in the southern part of the province.

It is also telling that the most significant new policy attributable to the drought was a soft policy rather than a hard one. In 2003, the Alberta government introduced its Water for Life program, a package of soft policy initiatives designed to enhance water conservation in the province, particularly in the south. Water for Life relied primarily on public education, moral suasion, and public spending as instruments to improve water conservation, and did not introduce any substantive changes to the water allocation rules in the SSRB. Consequently, the initiative has been criticized as toothless, but it is certainly soft in orientation, and it is notable that the province opted for this soft approach over the various hard policy measures that were also on the policy agenda at this time.

By 2000, it was clear that a path to major policy change had opened up in the Alberta water policy subsystem, presenting an opportunity to both advocacy coalitions. It started as an internal perturbation related to the planned year 2000 review of the SSRB Water Allocation Regulation, which morphed into a wider-ranging water management planning exercise after the Water Act took effect in 1999. The onset of the drought in 2000 and its persistence into 2001 constituted an external shock to the policy subsystem that added urgency to the planning exercise and resulted in its bifurcation. The drought also created an opportunity for the coalitions to advocate other water policy measures, including a number of hard projects and soft conservation measures.

In the remainder of this chapter, we will investigate all of these policy processes, starting with phase 1 of the SSRB Water Management Plan, then shifting focus to the proposed hard projects and the Water for Life initiative, before finishing with phase 2 of the SSRB Water Management Plan.

SSRB Water Management Plan: Phase 1

The SSRB water management planning process got underway in the second half of 2000 as Alberta Environment began inviting stakeholders to sit on the four basin advisory committees. This continued through the early part of 2001 but was interrupted by the March 2001 general election, which saw Ralph Klein and the PCs claim another resounding victory. By the summer, the rosters of the basin advisory committees were largely finalized and Alberta Environment began to host sessions to familiarize the participants with their tasks and objectives and to educate them on some SSRB water issues. The intention, at this point, was to produce a comprehensive water management plan that included water trading rules, conservation holdback rules, and water conservation objectives, all of which were called for in the new Water Act.

While the basin advisory committees and Alberta Environment were undertaking their planning preparations, the drought entered its second growing season, and the Southeast Alberta Water Co-op hatched its plan to buy water and pipe it to their parched corner of the province. With the co-op plan in mind, and with the threat of the drought stretching into a third growing season, Taylor viewed the introduction of water trading as a matter of urgency. He instructed his officials to expedite the introduction of water trading rules, through a water management plan, before the next irrigation season. To facilitate this, the planning process was divided into two phases, the first one dealing with water trading and conservation holdback rules, and the second one dealing with the more thorny and complicated issue of water conservation objectives.

The newly defined phase 1 of the planning process got underway in October 2001 and it took less than three months for a draft plan to take shape. During this time, the basin advisory committees met on a number of occasions, each time facilitated by officials from Alberta Environment. General agreement on a set of water trading and conservation holdback principles was found relatively quickly, and these formed the basis of the draft plan. That agreement was found so quickly is pretty remarkable, given the Aggie-Green dynamic on the basin advisory committees, but can be attributed to a combination of factors.

One key factor was policy-oriented learning in both coalitions. Between 1995 and 2002, many Aggies came to view conservation holdbacks as an acceptable water policy instrument. For Aggies, the issue of environmental water recoveries had long hinged on whether affected licence holders would be financially compensated for their losses. In 1995, as shown in table 4.2, the Aggies' stance on the issue was clear: they were opposed to any uncompensated environmental water recoveries but would accept some compensated water recoveries. The conservation holdback mechanism introduced by the Water Act was an instrument of uncompensated water recovery, much to the Aggies' dismay, but by 2002 they had largely withdrawn their opposition to it. Table 4.3 shows that no Aggies took a clear stand against uncompensated water recoveries in the phase 1 consultations. This is not to say that the Aggies were enthusiastic supporters of the conservation holdback mechanism, but they grudgingly accepted it because it was inextricably linked with water licence trading, which most Aggies were keen to see introduced. Aggies also came to learn that the amounts of water recovered through the conservation holdback mechanism would be relatively small, that there were some loopholes in the Irrigation Districts Act that would allow them to avoid holdbacks in some instances,[12] and that holdbacks would not be mandatory with every water trade, all of which helped in their acceptance of this previously unacceptable water policy instrument.

Similarly, between 1995 and 2002, many Greens came to accept water licence trading as a viable water policy instrument for the SSRB. Greens had long opposed water licence trading because they believed that it commodified an essential part of nature, opening up water resources for even further capitalistic exploitation. As shown in table 4.2, this opposition

12 In addition to the "ordinary" water rights that municipalities, irrigation districts, private irrigators, and others have under the Water Act, there are also derivative rights that individuals can obtain as a water user in an irrigation district. These derivative rights are regulated under the Irrigation Districts Act, which was amended to allow irrigators within an irrigation district to move water entitlements between parcels of land, subject only to the district's approval, and not subject to the requirements of the Water Act, which holds that transfers can be approved only if the director judges them to have no adverse effect on the aquatic environment (see Bankes and Kwasniak 2005, 6).

was clearly evident in 1995 with many Green groups taking a stand – unsuccessfully – against the introduction of water trading. By 2002, Green attitudes to water trading had warmed considerably. As table 4.3 shows, many Greens took no stand on water trading and some even expressed support for it. Much like their Aggie counterparts with respect to conservation holdbacks, most Greens interviewed for this study expressed a grudging acceptance of water trading rather than enthusiastic support. Green acceptance of water trading is attributable largely to its close linkage with conservation holdbacks, the only policy instrument available for reallocating any amount of licensed water for environmental protection purposes. It also helped that the Greens, during the course of the phase 1 negotiations, secured some water trading rules to ensure that water trades did not adversely affect riverine environments, which had always been their root concern about water trading.

So with the Aggies coming to accept conservation holdbacks and the Greens coming to accept water licence trading by 2002, there was a basis for general agreement on the main phase 1 planning issues. What remained was to work out the details of the water trading and conservation holdback rules, which was the main focus of the basin advisory committees during the three months it took to hammer out a draft plan.

Another factor facilitating the speedy development of the phase 1 draft plan was the design of the process itself. The basin advisory committees worked on a consensus basis, but they did not vote and unanimity was not required. When there were dissenting views to a general agreement, these views were noted in the reports to the Steering Committee, alongside the generally agreed recommendations. This meant that the actors with the most extreme views – in either coalition – could not block an emerging consensus by holding out. The few Greens who still vehemently opposed water trading and the few Aggies who still rejected conservation holdbacks could express their views, have their views duly noted, and attempt to persuade committee members to their way of thinking. But they had little success in doing so and had no procedural recourse for blocking the plan's development. This made it easier for most Aggies and Greens to find common ground, particularly in the context of their mutual policy-oriented learning.

Alberta Environment's brokerage role was also an important factor in the rapid development of the phase 1 plan. Officials from the department supported and facilitated the basin advisory committee meetings, helping them to reach consensus whenever possible. On the basis of committee reports and the requirements of the Water Act, the same officials then wrote the draft plan and organized public consultations on it during January 2002. The consultations were held at seven locations across the SSRB and, although some groups continued to object to water trading in principle, overall opposition to the draft plan was weak, as the core actors in both advocacy coalitions had already bought into it (Alberta Environment 2006, 28).

Near the end of January, Alberta Environment officials organized a plenary meeting of all the basin advisory committees to consider the public feedback and to address a few remaining points of contention. The notes from this meeting further indicate that there was broad agreement among committee participants on both water licence trading and conservation holdbacks, even if there was some remaining disagreement on specific water trading and conservation holdback rules (Alberta Environment, "Joint Meeting of the Basin Advisory Committees" 2002). After the plenary session, department officials put finishing touches on the plan and presented it to their minister. Most phase 1 participants interviewed for this study described it as a highly structured process from start to finish, closely guided by Alberta Environment officials determined to see it through to an expeditious conclusion.

Underlying the officials' determination was, of course, the strong political backing given to the completion of the phase 1 plan. From the time the planning process was divided into two phases, it was very clear to departmental officials and basin advisory committee members that the introduction of a (partial) basin plan to routinize water licence trading was a high priority for the government. This gave the phase 1 process a strong sense of inevitability, strengthening the hand of the officials trying to guide it, and convincing both Aggies and Greens that they could influence the process but not halt it. In other words, bargaining for the phase 1 plan took place in the shadow of considerable political determination to see a plan implemented, so both coalitions bargained to shape the plan as best they could, fearing that they might simply be bypassed if they held out for too much.

Overall, a combination of policy-oriented learning, committee decision-making by consensus (but not unanimity), coalition brokerage by Alberta Environment, and strong political backing resulted in the completion of a phase 1 basin plan in about nine months. Cabinet approved the plan in June 2002, just in time for the new irrigation season, and one of the first water licence trades approved under the plan was a purchase by the Southeast Alberta Water Co-op. Nearly all participants and observers of the phase 1 process agree that developing the plan in

Table 8.1 Conditions for Evaluating Proposed Water Trades

- There can be no significant adverse effect on the aquatic environment and no significant adverse effect on water conservation objectives.
- There can be no impairment of other licensees or rights holders (unless they agree to the impairment in writing).
- If the transfer is for irrigation purposes, the land to be irrigated must be suitable for irrigation.
- There can be no significant adverse effect on public health.
- There can be no significant adverse effect on groundwater quality or quantity.
- There can be no significant adverse effect on the operation of reservoirs or water infrastructure.
- There can be no adjustment to licence conditions, unless necessary to give effect to the transfer.
- The terms of the Master Agreement on Apportionment must be respected.

Source: Alberta Environment (*South Saskatchewan River Basin Water Management Plan* 2002, 10).

nine months was remarkably fast, certainly much faster than any previous water policy process analysed in this volume and faster than the phase 2 process, which spanned more than three years.

The design of the phase 1 plan clearly reflected negotiated compromises between the Aggie and Green coalitions. The grand compromise was the Green acceptance of water licence trading in return for the Aggie acceptance of conservation holdbacks, though smaller compromises are evident in the details of the plan as well. The plan authorizes the trading of water licences in all four sub-basins of the SSRB, though all water licence transfers must be approved by the plan's director, a senior official from Alberta Environment. In evaluating proposed licence transfers, the director must ensure that any transfer meets a lengthy list of conditions (summarized in table 8.1) before approval is granted. It is many of these conditions that were the subject of negotiation on the basin advisory committees and where Aggie-Green compromises were made. The plan also provides the director with the discretion to undertake conservation holdbacks: "The Director is hereby authorized to withhold up to 10 percent of an allocation of water under a licence that is being transferred, if the Director is of the opinion that withholding water is in the public interest to protect the aquatic environment or to implement a water conservation objective" (Alberta Environment, *South Saskatchewan River Basin Water Management Plan* 2002, 11). The Greens had pushed to make conservation holdbacks mandatory rather than discretionary, but the Aggies successfully resisted further encroachment of this power.

The introduction of the phase 1 water management plan marked a milestone in the evolution of water policy in southern Alberta. For the first time, water licence trades could be approved through a routinized administrative process, rather than having to go through the much more cumbersome process of Cabinet approval. Also unprecedented was the introduction of a mechanism to recover limited amounts of environmental water from traded licences, through conservation holdbacks. Both reforms were emblematic of the overall shift to softer water policy in the SSRB and heralded a new era in which water users would cope with scarcity through water redistribution and conservation rather than through new programs of dam construction and infrastructure expansion.

The Path Not Taken

The 2000–2 drought not only added urgency to the planned introduction of water licence trading, it also opened a path for other water policy reforms, both hard and soft. Historically, this has been a common pattern in southern Alberta and, indeed, in other water-scarce polities. Droughts serve as stark public reminders of the endemic water scarcity in a region, drawing media attention to current and forecasted water shortages, sparking constituent complaints to politicians about lack of water, and leaving politicians scrambling to somehow "drought-proof" afflicted areas. Recall, for instance, that the final decision to construct the Oldman Dam at the Three Rivers site in 1984 was precipitated by a drought and the Aggies' demands that the government do something to help them. The 2000–2 drought opened a path for the adoption of any water policy measures that would help southern Albertans cope with water scarcity and, in this way, provided an interesting test of the government's commitment to softer water policy. Would the government continue to introduce soft reforms, or would it revert to a program of hard infrastructure expansion?

There was certainly no shortage of hard projects that could be constructed, and some of these projects garnered support from key policy elites. As the drought intensified, Aggies dusted off and brought forward a number of hard projects, some of which had been on the shelf for over eighty years, including the construction of the Special Areas Water Project in the Red Deer sub-basin, the construction of a dam on the mainstem of the South Saskatchewan River, and the construction of an inter-basin diversion between the North and South Saskatchewan basins. While these projects did not attract wide public support, they were supported by key policy elites, particularly Environment Minister Lorne Taylor.

Taylor was one of the new breed of PC politicians who had entered politics during the Klein Revolution. He was part of the group of committed neoconservatives in the party caucus known as the "deep six," and his belief in the positive power of market forces was a key factor underlying his support for water licence trading. Somewhat paradoxically, Taylor was also a long-time proponent of large hard water projects, though this is largely explained by the fact that he represented the riding of Cypress–Medicine Hat, one of the driest in the province. During the legislative debate on the Water Act, for instance, Taylor was quite explicit about his support for further water infrastructure expansion:

> I think in some sense I would like to see the Bill go a little further. I think it makes it a little too difficult for us to get another dam in southern Alberta. We need a dam at Empress. It would be called the Meridian dam, or that's what the community group is calling it just for a token name. What this dam at Empress would do is back up the South Saskatchewan River and open up thousands and thousands of acres to irrigation. I think we need to be more open towards those kinds of issues in a Bill like this … I would [also] like to see the Bill actually encourage the development of a program of interbasin transfer, encourage at least some research. This would take a number of years to do this adequately, but if we don't start now, to start it 10 years from now is far too late, Mr Speaker. We need to start that process now. We need to be encouraging in this Bill a research approach to interbasin transfer, and I think it would certainly benefit the whole province. (Legislative Assembly of Alberta, 14 August 1996, 2165)

Taylor became environment minister in March 2001, as the drought entered its second year, giving the Aggies a well-placed proponent for their hard water projects and, seemingly, a good opportunity to capitalize on the water policy reform path that was just opening up.

The first hard project to come forward was the Special Areas Water Project in January 2001. Versions of this project had been floated since the 1920s, with the basic aim of irrigating vast tracts of land in east-central Alberta that were mostly unsuitable for dryland agriculture. In Alberta's early days, many of these lands had been occupied by dryland homesteaders, only to be abandoned during the 1920s and 1930s in a massive rural depopulation. Since then, the area had been so sparsely inhabited that it did not have the population base to support local government and was administered by the provincial Special Areas Board (Jones 1987). It was the Special Areas Board that, in January 2001, forwarded to Cabinet a $168 million proposal to divert water from the Red Deer River to support irrigation expansion in the Special Areas, which were suffering under the drought (Thomas 2001). Despite support from a number of regional MLAs, Cabinet opted against further study of the proposal and it went no further.

The next hard project to come forward was the Meridian Dam proposal, and it made it one step further than the Special Areas Water Project. Two months after he became environment minister, Taylor ordered a feasibility study of the proposed Meridian Dam, conducted jointly with the Saskatchewan government. This was the same project that Taylor had called for during the Water Act debates, and it involved the construction of a large dam on the mainstem of the South Saskatchewan River, near the Alberta-Saskatchewan border. The dam would create a reservoir about 150 kilometres long, bringing enough water under control to support an additional 160,000 hectares of irrigation, some of it in Taylor's own riding (Thomas 2001). The study, released in March 2002, estimated the cost of the dam to be $2.7 billion, much more than the economic benefits to be realized from the project, and not including the damage it would cause to the natural environment and to the interests of instream users such as boaters and anglers. The projections were so negative that even Aggies had difficulty putting a positive spin on them, and the project was quietly dropped by both governments (Canadian Broadcasting Corporation 2002; Lowry 2006, 326).

The most controversial hard project proposal during the drought of 2000–2 was the inter-basin diversion of water from the water-rich North Saskatchewan Basin to the water-poor South Saskatchewan Basin. As seen earlier in this study, inter-basin diversion proposals of this sort had existed for decades, the most prominent being the PRIME proposal developed in the 1960s. In December 2001, Environment Minister Taylor caused a political stir when he publicly mused about inter-basin diversions to the SSRB, even calling for a government study of the issue (Gillis 2001). Green backlash was swift and Premier Klein, a veteran of the Oldman Dam conflict and quite cognizant of Green soft power, moved quickly to quash the idea: "You run the danger of seriously disrupting the ecosystem. I'm of the mind that it ought not to be done. I'm telling you, if you want to create an environmental uproar in this province, go ahead with an interbasin transfer program" (Brooymans and Hryciuk 2002).

This effectively ended any serious consideration of inter-basin diversions by the Alberta government, for the time being. It was the third major hard water project to reach the policy agenda in 2001, none of them getting even close to adoption.

The only water infrastructure projects even partly attributable to the drought were a few relatively small ones funded through established infrastructure expansion programs. The best example was the South East Alberta Rural Trunk Water Line and Distribution System sought by the Southeast Alberta Water Co-op. This project was constructed through the Infrastructure Canada-Alberta Program on a one-third cost-sharing basis between the government of Alberta, the government of Canada, and the co-op itself. This program was administered through the Department of Transportation, and the pipeline proposal received approval for $25.4 million of funding in March 2004. This was the final piece in the plans of the Southeast Alberta Water Co-op, allowing them to purchase water from the United Irrigation District and transport it 200 kilometres east to co-op members in the counties of Lethbridge, Warner, Forty Mile, and Cypress (Canadian Consulting Engineer 2004; Legislative Assembly of Alberta 2004, 845).

The policy reform path opened by the drought created an opportunity for the introduction of soft water policy measures as well as hard ones, and the soft policy measures fared much better. This was most evident in November 2003 when Alberta Environment launched its Water for Life initiative. In many ways, Water for Life complemented the ongoing water management planning process in the SSRB. It was designed as a bundle of soft policy measures to foster a new approach to water management in the province, one relying on conservation more than construction. It included new funding programs to support research and public education on water issues, and it called for the introduction of integrated watershed management planning throughout the province (Alberta Environment 2011).

Water for Life's integrated watershed management planning initiative was different from the SSRB planning process, though both were soft. While the SSRB Water Management Plan was mostly about water allocation between competing uses, the integrated watershed management plans were more holistic, addressing water quantity and quality issues together. To facilitate the new watershed management planning initiative, Water for Life created a three-tiered institutional structure, including a new Alberta Water Council at the provincial level, various Watershed Planning and Advisory Councils at the basin/sub-basin level, and many Watershed Stewardship Groups at the local level. Representatives from various stakeholder groups, including Aggies and Greens, as well as members of the interested public, were invited to sit on these planning groups, and a number of them have made strides in developing integrated watershed management plans (Berzins, Harrison, and Watson 2006). The Oldman and Bow sub-basins, for instance, have particularly active watershed planning and advisory councils, each of which has developed integrated watershed management plans in the past few years.[13]

Water for Life also involved a fairly extensive public consultation process in March–April 2002. One part of this was a random telephone survey of 1,000 Albertans which, overall, showed considerable public support for the Greens' water policy preferences. For example, 87 per cent of respondents agreed or strongly agreed that "if there are water shortages in the future, the Province should put a higher priority on preserving natural aquatic environments, even if this limits economic growth and jobs" (Equus Consulting Group Inc. 2002, 18). Eighty-one per cent also agreed or strongly agreed that the government needs to "maintain an amount of water in aquatic environments that will protect ecosystems even though it may limit human use" (18). The survey further showed majority public support for water licence trading and increased water pricing as means for improving water use efficiency. There was also majority public support for the investigation of new storage reservoir sites, but strong public opposition to inter-basin diversions (10–11). Thus, public opinion was solidly behind the Greens on many water issues, giving them a very important soft power resource both in the development of Water for Life and in the negotiation of the SSRB Water Management Plan.

While the scale of the Water for Life initiative is impressive, and it has been lauded as a participatory and progressive approach to water governance, its impact has been less consequential than often claimed and certainly far less consequential than the SSRB Water Management Plan. Water for Life's integrated watershed management plans

13 These watershed planning and advisory councils are distinct from the basin advisory committees involved in the SSRB water management planning process, though some individuals have served on both.

are non-binding, effectively constituting a set of recommendations to the ministers involved in the governance of a river. Whereas the SSRB Water Management Plan fundamentally changed the way in which water could be allocated and reallocated within the basin, the effects of the integrated watershed management plans have been marginal. The integrated watershed management plans may come to take on additional importance over time, but they have been far less impactful than the SSRB Water Management Plan.

Despite the weaknesses of the Water for Life initiative, its adoption was still important, particularly as an indicator of how much things had changed in the Alberta water policy subsystem. In the midst of a severe drought, the government of Alberta opted for the soft approach of Water for Life over the hard approach embodied in the Special Areas Water Project, the Meridian Dam, and in inter-basin diversions. And in making these decisions, policymakers explicitly acknowledged the influence of environmental considerations and, even more pointedly, the political influence of Greens. The Alberta water policy subsystem was now solidly characterized by Aggie-Green pluralism, and Green influence prevented policymakers from falling back into their old hard policy habits, even with a drought-initiated policy reform path at their disposal.

SSRB Water Management Plan: Phase 2

Shortly after Cabinet approved the phase 1 SSRB Water Management Plan in June 2002, preparations for phase 2 got underway.

Phase 2 promised to be a more complicated process than phase 1. The focus of phase 2 was the development of water conservation objectives, which, in effect, involved defining what shares of water would be set aside for the environment and what shares of water would remain available for licensed users. These were complex issues involving assessments of environmental water needs, users' water needs, and total water availability. Moreover, these factors varied between subbasins, between rivers, and even between reaches of the same river, requiring a lot of information and localized knowledge in order to formulate proper water conservation objectives. There was also some disagreement, even among scientists, about the actual water needs of the environment and of water users, and there was some uncertainty about total water availability in the changing climate. Such complexity was problematic because it left considerable room for the advocacy coalitions to try to frame the issues in a self-serving manner, making it difficult to find common ground between them.

The Aggies and Greens also seemed to be further apart on water conservation objectives than they had been on water licence trading and conservation holdbacks in phase 1. Aggies feared that stringent water conservation objectives would become a back door means of constraining their water use, thereby undermining their water licences. In a survey of irrigation district managers and board members, 63 per cent of respondents identified "increasing environmental water requirements" as a high or very high threat to irrigation water availability. This was the most frequently identified threat to irrigation water, more than global warming, the Canada-U.S. apportionment agreement, increasing livestock water requirements, increasing city and municipal water requirements, or increasing industrial water requirements (Bjornlund, Nicol, and Klein 2006, 6). For the Greens, phase 2 was crucial because it would finally put into action a key eco-protection policy instrument they had fought to include in the Water Act. If water conservation objectives could be established to protect instream flows, then the decline of riverine environments in the SSRB could be halted and eventually reversed. In short, the phase 2 negotiations had a zero sum dynamic between Aggies and Greens that the phase 1 negotiations had not, promising to make the development of the phase 2 plan much more difficult.

Alberta Environment officials recognized the considerable challenges of the phase 2 negotiations and, in their brokerage role, took measures to mitigate and manage them. Between July 2002 and June 2003, the department prepared an extensive set of background studies, listed in table 8.2. Some studies were written by departmental officials while others were written by outside consultants or academics, but all of them were based on the best available science (Alberta Environment 2005, 36–7). The importance of the background studies can hardly be overstated. They helped frame the phase 2 issues in empirical terms, reducing some of the endemic complexity and uncertainty, and limiting the coalitions' ability to frame issues in ideological terms. For the Greens, the background studies were also a newfound source of soft power. A number of the studies lent support to Green policy beliefs, and the Greens repeatedly pointed to them in their efforts to draw other stakeholders towards their policy positions.

By June 2003, the background studies were complete, the basin advisory committees were reconstituted, and the committee members were given copies of the background studies to peruse over the summer. The four basin advisory committees from phase 1 were retained, though

Table 8.2 Background Studies for the Phase 2 SSRB Water Management Plan

Studies of water demand and consumption	• "South Saskatchewan River Sub-Basin Contributions to International and Interprovincial Water-Sharing Agreements" (by Alberta Environment) • "South Saskatchewan River Basin Water Allocation" (by Alberta Environment) • "South Saskatchewan River Basin Non-Irrigation Water Use Forecasts" (by Hydroconsult EN3 Services Ltd)
Studies of the state of the aquatic environment	• "Instream Flow Needs Determinations for the South Saskatchewan River Basin" (by Clipperton et al.) • "Report on Strategic Overview of Riparian and Aquatic Condition of the South Saskatchewan River Basin" (by Golder Associates)
Water modelling	• "South Saskatchewan River Basin, Water Management Plan, Phase 2, Scenario Modelling Results, Part 1" (by Alberta Environment)
Studies of recreational water use	• "Recreation Flows for the Bow River and Its Tributaries" (by Tymensen and Rood) • "Recreation Flows for Paddling along Rivers in Southern Alberta" (by Rood and Tymensen) • "Recreation Flows for the Red Deer River" (by Rood and Tymensen)

there was some turnover in committee membership. The consensus decision-making rule on the committees was also retained and their mandates were largely the same: to make recommendations to the Steering Committee on draft water management plans for their respective sub-basins. The most notable procedural change from phase 1 was Alberta Environment's employment of Equus Consulting as a committee facilitator. The facilitator met regularly with the basin advisory committees, advising them on technical issues and helping them formulate their recommendations (Alberta Environment 2005, 27). In the background was Alberta Environment, overseeing the committees' operations and offering advice whenever requested.

The phase 2 negotiations began in earnest in the fall of 2003 and the first issue addressed by the basin advisory committees was the closure of the Bow, Oldman, and South Saskatchewan sub-basins to new water licences. This was an issue that had come up in the phase 1 negotiations but had been punted to phase 2. Parts of the Oldman sub-basin had been administratively closed to new water licences since the late 1970s, but the new Water Act created the power to impose licensing moratoria as a matter of policy. Hence, the issue was whether to impose licensing moratoria in one or more of the heavily allocated southern sub-basins where most of the irrigation in the province took place.

Early in the phase 2 negotiations, a remarkable degree of consensus emerged in favour of imposing moratoria in all three southern sub-basins. The Greens had called for licensing moratoria for years, so their support was not at all surprising. Much more unexpected was the support of Alberta Agriculture, the Alberta Irrigation Projects Association, and most of the irrigation districts, all of them central figures in the Aggie coalition. Not all Aggies supported the moratoria, but enough of them supported it that working consensuses were reached on the Bow, Oldman, and South Saskatchewan basin advisory committees.

The Aggies' newfound support of licensing moratoria can be explained by two main factors.

The first factor was the background studies on water demand and consumption and the state of the aquatic environment. Almost all of the phase 2 participants interviewed for this study – whether Aggie or Green, governmental or non-governmental – pointed to these background studies as crucial in persuading Aggies that the three southern sub-basins should be closed to new water licences. The studies showed that the water supply and demand curves were rapidly approaching each other and that current flows were well below the minimum levels for instream flow needs. The data were so compelling and so difficult to refute that the Aggies were ultimately persuaded of the necessity of basin closure.

The second factor was, quite simply, that the moratoria would affect prospective irrigators (without licences) but not existing irrigators (who already had licences), and it was existing irrigators who were sitting at the bargaining table. Some Aggies also calculated that allowing the rivers to become increasingly over-allocated could, in the future, spark a backlash against all water licence holders, putting all irrigators' entitlements at risk. In this line of thinking, it was better to accept the moratoria now than to have licence cutbacks later.

With the moratoria settled in relatively short order, the basin advisory committees turned their attention to the more controversial issue of water conservation objectives and the share of water that should be set aside for environmental use. The issue turned largely on varying perceptions and definitions of instream flow needs. The background studies provided a definition of instream flow needs, recommending that around 85 per cent of natural flows be reserved to support fully functioning natural river ecosystems (Pentney and Ohrn 2008, 391). The Greens adopted this as their preferred standard, arguing strenuously that their position was scientifically based. For the Aggies, this

85 per cent figure was far too high. They argued, instead, that Alberta Environment's current "instream objectives" should serve as the basis for the new water conservation objectives. The instream objectives had existed as flow targets since the 1980s, though they had no legislative or regulatory basis and could not be enforced. The instream objectives would have reserved less than 20 per cent of natural flows on some rivers, and the Aggies argued that these levels had been determined by experts as sufficient to sustain fish and other aquatic species. This was a considerable distance from the Green position, and the result was Aggie-Green splits on the basin advisory committees.

Although it was the least heavily allocated of the four sub-basins, the development of water conservation objectives was most controversial in the Red Deer sub-basin. The three southern sub-basins were already so heavily allocated with existing licences that, in most years, the remaining flows would not even come close to meeting the low-end water conservation objectives favoured by the Aggies. Since existing water licences were untouchable under the Water Act, the water conservation objectives in these sub-basins were largely aspirational, constituting a "soft cap" that would be frequently exceeded. The Red Deer, however, was not nearly as heavily allocated and, in most years, there was enough water in the river to meet the high-end water conservation objectives favoured by the Greens. Once introduced, the water conservation objectives in this sub-basin would constitute a "hard cap," as the share of water set aside for the environment would be inviolable and unavailable for future allocation. In this way, the water conservation objectives in the Red Deer were effectively establishing a ceiling on new water licences in this sub-basin, significantly shaping the future course of water development in the area. This raised the stakes on the Red Deer Basin Advisory Committee, and it was there that the main battle over water conservation objectives was fought.

To overcome the Aggie-Green split on water conservation objectives, the Red Deer committee turned to the Steering Committee for help. The Steering Committee, in turn, asked the working group that had put together the background studies on instream flow needs to provide more information on various water conservation scenarios. The result was the creation of the so-called impact stick, "a scale illustrating potential impacts on the aquatic environment (a composite of riparian vegetation, fish habitat, water quality and channel maintenance) of different river flow regimes that would result from varying degrees of allocation" (Pentney and Ohrn 2008, 392). The impact stick showed that reducing the water conservation objective from 85 to 45 per cent of natural flows had only a marginal impact on the aquatic environment; 45 per cent had more impact than 85 per cent, but not much more (ibid.). This finding undercut the Greens' arguments and diminished their soft power on the issue. The Steering Committee, the Aggies, most of the unaligned stakeholders, and even some moderate Greens adopted the 45 per cent figure as their preferred water conservation objective, and a working consensus emerged around it. Many Greens continued to argue for a higher objective, but to little avail.

The three southern basin advisory committees had watched these events transpire and were strongly influenced by the Red Deer outcome. The 45 per cent figure became the basis of working consensuses on each of the Bow, Oldman, and South Saskatchewan committees, even though there was as little as 5 or 10 per cent of natural flows actually available in some rivers once all existing water licences were met. In these sub-basins, the 45 per cent figure was largely aspirational, though it did signify that "if any opportunity arises to return water to these rivers, it should be taken" (Pentney and Ohrn 2008, 388). This would encourage the taking of conservation holdbacks from water trades in the southern sub-basins, thereby slowly – very slowly, given the small amounts of water held back – returning water to the rivers and moving towards the 45 per cent objective. For Greens, the 45 per cent figure also had symbolic value as it was an implicit admission that the southern sub-basins were grossly over-allocated, perhaps justifying a larger effort to return more water to the rivers at some point in the future.

Through most of 2005, the Steering Committee members engaged in extensive shuttle diplomacy as the phase 2 plan slowly took shape. From January through July, they shuttled between the basin advisory committees, the various government departments involved in the plan, and Environment Minister Guy Boutillier, who would ultimately bring the plan forward to Cabinet (Ohrn 2005; "Working Outline: SSRB Water Management Plan" 2005). In April and May, a separate one-on-one consultation process with the Treaty #7 First Nations, who had declined to participate on the basin advisory committees, was undertaken by Alberta Environment officials (Alberta Environment 2005, 27). Internal emails and memos also show that the Steering Committee was quite cognizant of public opinion in developing the draft plan and one of their chief objectives was to create a plan that would be publicly acceptable. By the end of July, the main planning issues had been settled and the Steering Committee set about writing a draft plan. The draft plan was released on 18 October and the public consultation process began (ibid.).

The public consultations mirrored the basin advisory committee negotiations in the sense that there was strong support for the licensing moratoria, but mixed support for the water conservation objectives. The public consultations took place between October 2005 and January 2006, with seven public meetings held at various locations throughout the SSRB. Surveys showed that 81 per cent of meeting participants supported the closure of the Bow, Oldman, and South Saskatchewan sub-basins to new licences, with only 11 per cent opposed (Equus Consulting Group Inc. 2006, 7). However, support for the proposed water conservation objectives was divided almost evenly among those who thought they were acceptable, those who thought they were unacceptable, and those who were unsure. Most of those who found them unacceptable opposed them on the grounds that too little water had been set aside for instream flow needs (9–11). With public opinion divided, neither advocacy coalition was able to capitalize on public support to push for their preferred lower/higher water conservation objectives, so the 45 per cent figure was retained as a compromise measure.

Over the next four months, the Steering Committee revised the draft plan in light of public input and final comments from the basin advisory committees. In May, a complete plan was then presented to Environment Minister Boutillier, who endorsed it and presented it to Cabinet in the summer. In August, the phase 2 plan was approved by Cabinet, who formally adopted it as an amendment to the already existing (phase 1) SSSRB Water Management Plan from 2002. The water conservation objectives were then implemented in January 2007 through directives from the Water Act's designated director, and the licensing moratoria were implemented in August 2007 through Regulation 171/2007, pursuant to the Water Act.

Although the details of the water conservation objectives vary between sub-basins and between the upstream and downstream portions of each sub-basin, the target of 45 per cent of natural flows is central to all of them. The water conservation objectives for the Bow sub-basin are illustrative. Downstream of Bearspaw Dam, the water conservation objective for the mainstem of the Bow is "either 45% of the natural rate of flow, or the existing instream objective increased by 10%, whichever is greater at any point in time" (Alberta Environment, "Establishment of Bow River Sub-Basin" 2007, 1). Upstream of Bearspaw Dam, the water conservation objective for the Bow and its tributaries is "not less than the existing instream objective or the [water conservation objective] downstream on the mainstem, whichever is greater at any point in time" (1). The other three sub-basins have similar formulations, with upstream and downstream designations that make sense based on the hydrology of each sub-basin (Alberta Environment, "Establishment of Oldman River Sub-Basin" 2007; Alberta Environment, "Establishment of Red Deer River Sub-Basin" 2007; Alberta Environment, "Establishment of South Saskatchewan River Sub-Basin"). However, it is important to emphasize again that the water conservation objectives for the Bow, Oldman, and South Saskatchewan constitute aspirational "soft" caps on water use, while the water conservation objectives for the Red Deer constitute a substantive "hard" cap on (future) water use.

To give effect to the licensing moratoria in the Bow, Oldman, and South Saskatchewan sub-basins, Regulation 171/2007 introduced a Crown reservation of all unallocated water in these sub-basins. It further specified that water licences for this unallocated water could be granted only for very specific purposes, one of the most prominent being the achievement of water conservation objectives. However, it also allowed for the grandfathering of some non-environmental water uses, including water licences for First Nations when their outstanding water rights claims are settled; water licences for improving water storage, if these licences improve water availability and benefit the aquatic environment; water licences for applicants with pending water licence applications at the time the regulation was adopted; and water licences related to two small dam projects in the Oldman sub-basin that had already been approved and were being built at the time the regulation was adopted (Province of Alberta 2007). So the moratoria had some holes in them, but the holes were fairly small and most would close with the passage of time. The new Crown reservation also superseded the Crown reservation from the 1991 SSRB Water Allocation Regulation, which had reserved all unallocated water for economic purposes and had been intended to support further irrigation growth. In this way, Regulation 171/2007 marked a significant reversal in southern Alberta water policy, going from the reservation of most unallocated water for irrigation to the reservation of most unallocated water for the environment.

Summary

With the conclusion of the SSRB Water Management Plan, the soft policy instruments created by the Water Act were put into effect and water governance in southern Alberta became markedly softer. Central to this policy shift was the presence of the Greens and their effective use of soft power in the plan's development. The Greens used favourable public opinion and scientifically based information to draw the Aggies and policy decision-makers towards some of their preferred policy options such as conservation holdbacks and licensing moratoria. The Aggies, however, remained the dominant coalition and were able to introduce water licence trading and protect their water licences by minimizing the impacts of the new water conservation objectives. Throughout the planning process, Alberta Environment played a key policy brokerage role, framing issues in empirical terms and building bridges between the coalitions. The resulting plan was very much a compromise between Aggie and Green policy preferences, and Alberta Environment deserves credit for helping the advocacy coalitions reach this common ground. Some of the effects of the SSRB Water Management Plan – and indeed the larger shift to softer water policy – will be explored in the first part of chapter 9.

Chapter Nine

Conclusion

Anyone who can solve the problems of water will be worthy of two Nobel prizes – one for peace and one for science.

John F. Kennedy

Over the last eight chapters, we have seen how southern Alberta has shifted to softer water policy in the last twenty years, and, using an ACF approach, we have used two central hypotheses to explain this policy shift. In this final chapter we will look at what has transpired in the southern Alberta water policy subsystem since the shift to softer policy, and make some projections about the future of southern Alberta water policy based on the policymaking dynamics identified in this book. We will also consider the study's findings in light of the ongoing development of the ACF as a theory of the policy process. The ACF is periodically revised and updated on the basis of results from empirical testing of its key propositions, and this study can make a useful contribution as the latest of these empirical tests. Finally, we will draw out some of the implications of the southern Alberta water case for the governance of wicked problems more generally, paying particular attention to the inherently and intensely political nature of wicked problems.

Since the SSRB Water Management Plan

As expected – and intended – the shift to softer water policy has fundamentally altered the political economy of water use in southern Alberta. Though water had always been scarce in the SSRB, the licensing moratoria have now made water *licences* scarce, changing the incentives for existing and prospective water users alike. Water licences are now valuable assets, and existing licence holders have stronger incentives to conserve and use their water more efficiently. In some cases, this means selling surplus water for profit, and in other cases it means selling entire allocations, as the licences themselves have become more valuable than the economic activity they supported. Prospective water users now turn to the market, rather than government, to find new water to support commercial and residential developments, and the costs of finding and purchasing water are now part of the cost of doing business. Most water licence trades have also been subjected to conservation holdbacks, slowly – very slowly – returning more water to the rivers in support of instream flow needs. Overall, whereas the political economy of water use for much of the twentieth century could best be described as "build and allocate," it is now very much one of "cap and trade."

The licensing moratoria in the Bow, Oldman, and South Saskatchewan sub-basins have largely had their intended effect of curtailing the growth of licensed water allocations. As figure 3.1 shows, the total volume of allocated water licences in the SSRB has nearly levelled off over the past decade. Any remaining growth has been due to the distribution of new licences in the Red Deer sub-basin where there is no moratorium, and to the granting of new licences in the southern sub-basins to those who were grandfathered into the moratoria there. Examples of the latter include some licences granted in relation to the Little Bow and Willow Creek projects completed in the mid-2000s, and a licence granted to the Piikani First Nation when they reached a water rights settlement with the Alberta government in 2010. Most of the grandfathered water users have now been licensed, so these loopholes are almost closed, and most future growth will come only from new licences in the Red Deer sub-basin.

The effects of water licence trading are somewhat mixed and difficult to evaluate, given that the water licence market is little more than a decade old. However, some inferences can be drawn by analysing the volume and patterns of water trading thus far, and the evidence seems to indicate that the water licence market is working as intended, even if water trading uptake has been relatively low.

Between mid-2003, when water trading first came into effect, and the end of 2014, a total of 163 water licence trades took place, collectively amounting to just under 33 million cubic metres of water. To put this in some context, the total volume of traded water constitutes just 0.00065 per cent of the total licensed volume in the SSRB, or about 7 per cent of the storage capacity of the Oldman Dam. Though one would not expect (or want) a large percentage of the available water to be traded, these figures do seem quite small and suggest that water trading uptake has been relatively low. While this may be disappointing to the most enthusiastic water market proponents, there is some evidence that the southern Alberta water market has been on the upswing in recent years.

It seems that the water market sputtered in its first few years but has expanded since. Figure 9.1 shows the number of water trades, the volume of water trades, and the volume of conservation holdbacks between 2003 and 2014. During the first six years of the water market, the number of water licence trades, with the exception of 2006, was quite low. Since then, the number of trades has been consistently higher, with the highest number occurring in 2012 and 2014, and the largest volume of trades occurring in 2013 and 2014. So it seems that water trading has picked up over the last six years and that we may be at the start of a longer-term upward trend.

All of this suggests that the first six years of the water market – and perhaps longer – were an awkward transition phase, as water users and regulators adapted to the new political economy of water use in the SSRB. This awkward transition phase may have occurred for number of reasons. First, it may have taken some time for water shortages to develop that could not be met with existing licences, thereby forcing water users into the water market. Second, it may have taken awhile for existing and prospective water users to become aware of the water market and to become comfortable with it. The buying and selling of water licences was a huge cultural change for water users in the SSRB, and big cultural changes are rarely embraced overnight. Third, the prospect of losing 10 per cent of a licence through a conservation holdback may have put an initial chill on trades, which later thawed as water users got used to the idea and were forced into the market out of necessity. Fourth, it may be that the transactions costs, in terms of finding willing buyers or sellers, were quite high in the early days of the water market, but have diminished over time as water licence transactions have been routinized. Whatever the reasons for the awkward transition phase, a more mature water market now seems to be emerging, one that is embraced and used by a wide variety of water users in the SSRB.

Figure 9.1 Water Licence Trades and Conservation Holdbacks in the SSRB (2003–14)

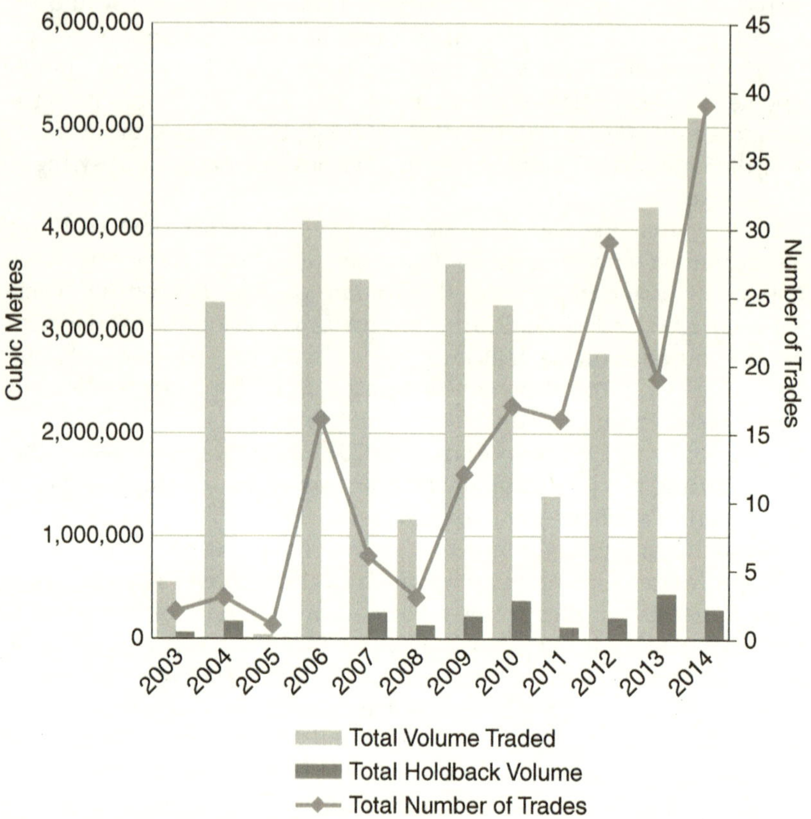

The position of the irrigation districts in the water market is particularly interesting and distinctive from other water users. Remember that the irrigation districts are the holders of most of the largest and oldest water entitlements in the SSRB, so any well-functioning water market would need irrigation district participation. The data in this regard are mixed: of the 163 total water trades, irrigation districts have been a party in only 5 of them; but these 5 trades have involved about 5.5 million cubic metres of water, accounting for almost 17 per cent of all water traded since the water market's inception. Moreover, a number of these trades could be characterized as important, high-profile trades, such as the trade between the Eastern Irrigation District and the Municipal District of Rocky View (recounted in chapter 1) and between the United Irrigation District and the Southeast Alberta Water Co-op (recounted in chapter 8). So, while the irrigation districts have not been perennially active in the water market, they have made quite a splash in the few instances when they have made water trades.

One reason for the irrigation districts' limited involvement in the water market is the existence of something called "amendment transfers," which allow irrigation districts to move water between different uses outside of the water market. A provision in the Irrigation Districts Act allows the districts to use irrigation water for non-irrigation purposes, as long as their licence allows it. So, as Kwasniak (2006) describes, many of the irrigation districts have simply sought licence amendments rather than use the water market: "There normally is more incentive for districts to apply to the province to amend their licences to authorize a variety of uses, so that they can annually charge new users, rather than forever losing allocations through transfers. Many new users in the SSRB will fall within one of the 13 irrigation districts. If they can get water from a district they need not seek water through transfers" (n95).

These amendment transfers are opposed by the Greens, as they allow irrigation districts to skirt the water market, avoid the conservation holdback mechanism, and retain their positions as the water barons of southern Alberta (Bankes and Kwasniak 2006). The amounts of water involved are also quite substantial. In 2003, the St Mary Irrigation District had its licence amended so that it could reallocate up to 15 million cubic metres – almost half the total volume traded in the water market thus far – for non-irrigation uses within its service area. The only times the districts have resorted to the market were when they were selling to someone at a distance, such as the two high-profile trades noted above, and the water was moving well beyond a district's service area. While amendment transfers do allow for some flexibility in shifting irrigation district water to new residential and commercial uses, they also allow the districts to bypass the water market, and, in this way, explain some of the limited uptake of water trading, thus far.

Figure 9.1 also shows that the number and volume of water licences traded in the SSRB have varied considerably from year to year; however, the reasons for this variance are not clear. Some of the variance between the early years and the later years may be accounted for by the awkward transition phase hypothesis described above.

Another factor that one would expect to affect water trading decisions would be weather, specifically annual variations in precipitation, but there does not appear to be a straightforward relationship between natural water availability and water trades. There were, for example, an almost equal number of trades in 2010 and 2013, though 2010 was a very dry year and 2013 was a very wet year. It may be that the fluctuation in water trades is due to a confluence of factors, some related to weather and climate, and some related to long-term demographic and economic structural changes in the region, factors that will become clear only with time.

In terms of the movement of water between water uses, the evidence generally indicates that the water market is functioning as would be expected. Figure 9.2 shows the movement of water (by volume) between various uses in the SSRB, based on Alberta Environment's classifications of the buyers and sellers[14] in the 163 water trades thus far. Water markets are generally expected to move water from lower-valued uses to higher-valued uses, which usually means that water moves from cropping (with lower rates of return per unit of water) to domestic and commercial uses (with higher rates of return per unit of water). This expected pattern has been evident in the SSRB: 27 per cent of the water traded has moved from irrigation to domestic/commercial uses, but there have been no instances of the reverse trading relationship. It is also noteworthy that, in all five trades involving irrigation districts, the districts have been sellers and the water moved from irrigation to domestic/commercial uses.

Nevertheless, most of the water traded in the SSRB has moved between irrigation uses or between domestic/commercial uses. Many of these trades probably also moved water from lower-valued to higher-valued uses within these use categories, but it is impossible to be certain without finer-grained data. Overall, the direction of water movement in the market seems to be consistent with the expectations of economic theory and with the experiences of longer-standing water markets in places like Australia and the western United States.

14 The methodology used by Alberta Environment in classifying water users was not specified and it is therefore impossible to evaluate how rigorous it is. However, their knowledge of the buyers and sellers and the purposes for which traded water was and will be used far exceeds mine, so I chose to rely on it as the best source available. The only adjustment I made to the data was to combine domestic, commercial, and municipal uses into a single category (domestic/commercial water use).

Figure 9.2 Movement of Water (by Volume) through Water Trades in the SSRB (2003–14)

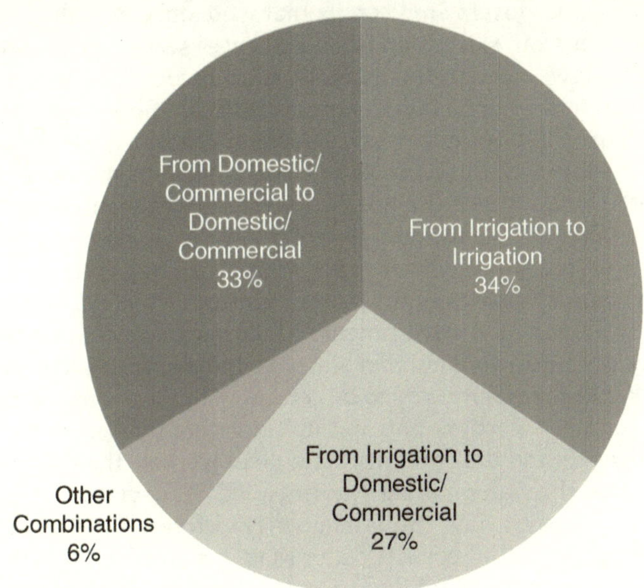

The data also show that the water market has been much more active in the southern part of the SSRB than in the northern part. For administrative purposes, Alberta Environment has divided the SSRB into a northern region (roughly encompassing the Red Deer and Bow sub-basins) and a southern region (roughly encompassing the Oldman and South Saskatchewan sub-basins), and the data show that the volume of water traded in the southern region has been more than twice that traded in the northern region. Given that water is generally scarcer in the southern region, this is an interesting result. It suggests that water users have turned to the market to help them cope with scarcity and, in this respect, the market is functioning as it should. In any water market in a situation of scarcity, there is always a danger that licence holders will sit on their licences and refuse to trade them for fear of losing their water and never getting it back. This does not appear to be happening in southern Alberta.

One thing that is not clear from the data is how many of the traded licences have been "sleepers" and "dozers," activated only by the introduction of the water market. Sleepers are water licences that existed prior to the introduction of the water market but not actually used by the holders, and dozers are licences that were only partially used. The unused water from sleeper and dozer licences served as de facto environmental flows prior to the market's introduction, as this water was left in the rivers where it could support instream flow needs. With the introduction of the water market, sleeper and dozer licences suddenly had a monetary value and could be traded to those who needed water and would actually use the licensed water rather than leave it in the rivers. In this way, the introduction of the water market has the potential of increasing water withdrawals, thereby leaving riverine environments worse off than before the market's introduction. Alberta has long had rules allowing the government to take back licences that were not being used, so it is unlikely that sleepers and dozers were rampant in the SSRB, but there is no way to be certain with the available data. The issue is certainly worth noting and investigating further, as the activation of sleepers and dozers could actually exacerbate the water scarcity problem that the water market was intended to alleviate.

One of the Greens' most important policy victories in the development of the Water Act and the SSRB Water Management Plan was the introduction of conservation holdbacks, the first and only policy instrument allowing for the reallocation of water from licences to the environment. This instrument allows for a holdback of up to 10 per cent of the volume of any water licence traded in the SSRB. Conservation holdbacks are at the discretion of the Water Act's director, and the water is typically reallocated to contribute to the water conservation objectives specified in the SSRB Water Management Plan.

Thus far, the amount of environmental water recovered through conservation holdbacks has been relatively small. Conservation holdbacks have been taken in 72 per cent of water trades, and in every holdback case the maximum 10 per cent has been taken. The main criterion for taking a holdback seems to be whether the director feels that a holdback would contribute to a water conservation objective within the vicinity of a trade. The total amount of water recovered through conservation holdbacks has been 2.23 million cubic metres, which is equivalent to 0.00004 per cent of all licensed allocations in the SSRB or about 0.5 per cent of the Oldman Dam's storage capacity. This is a small amount of water, but it was anticipated that the volumes of water involved would be relatively small when the conservation holdback instrument was being developed. However, the volume of holdback water will only grow over time, and the notion of reallocating some licensed water to the environment has been enshrined in policy, changing the starting point for future policy negotiations.

Much of the Greens' objection to the amendment transfers pursued by the irrigation districts has been their avoidance of conservation holdbacks. When irrigation districts move water between uses by amending their licences, no water trade takes place, and no conservation holdback can be taken. The Greens have argued that the irrigation districts should be forced into the water market in these cases so that holdbacks can be applied (Bankes and Kwasniak 2005). The conflict over amendment transfers came to a head in 2006–8 when the Eastern Irrigation District applied to have its licence amended to allow it to use a large volume of irrigation water for non-irrigation purposes. The Greens became aware of the application and lobbied hard against it. Eventually, the application was rejected and Alberta Environment announced a new policy whereby it would only consider amendment transfers of no more than 2 per cent of an irrigation district's licensed water allocation. This was a considerable victory for the Greens and for the nascent water market that, ironically, they had staunchly opposed when water trading was first proposed in the development of the Water Act.

The end goal of the conservation holdbacks is to provide enough environmental water to meet the water conservation objectives specified in the SSRB Water Management Plan, but outside of the Red Deer sub-basin, this goal remains largely unmet. The Red Deer is the least heavily allocated of the SSRB's sub-basins, and in most years there is enough water left in its rivers to meet its water conservation objectives. As new licences are distributed and increasing amounts of water are withdrawn in the future, the limits imposed by the Red Deer's water conservation objectives will eventually kick in, imposing a "hard cap" on licences so that the water conservation objectives will continue to be met. In the southern sub-basins, the situation is very different. These sub-basins are already so heavily allocated that the water conservation objectives are unmet in most years and constitute little more than aspirational "soft caps" on licences. The water recovered through conservation holdbacks is intended to gradually ratchet up environmental flows in the southern sub-basins towards their water conservation objectives, but because the amounts recovered through holdbacks have been small, so has been the progress towards achieving the water conservation objectives.

As far as the aquatic environment is concerned, it is not clear whether the shift to softer policy has resulted in a net gain or loss of water for the environment. This could probably be determined by comparing the amount of water recovered through holdbacks with the amount of water withdrawn through the activation of sleepers and dozers. Unfortunately, such data are not available, but this could be the subject of future study.

Since the conclusion of the SSRB Water Management Plan, the southern Alberta water policy subsystem has remained quite active, but no major water policy reforms have been pursued. The Water for Life program was renewed in 2009, and much attention in the subsystem has shifted to the integrated watershed management planning called for in this initiative. Watershed Planning and Advisory Councils have been created for each sub-basin of the SSRB: the Red Deer Watershed Alliance, the Bow River Basin Council, the Oldman Watershed Council, and the Southeast Alberta Watershed Alliance (Alberta WPACs 2015). Both Aggies and Greens have been well-represented on these councils, and watershed management planning efforts are ongoing in all of them. However, thus far, only the Bow and Oldman councils have produced integrated watershed management plans (Bow River Basin Council 2015; Oldman Watershed Council 2015; Red Deer Watershed Alliance 2015; Southeast Alberta Watershed Alliance 2015).

It is important to remember that these plans constitute a set of recommendations to the ministers involved in the governance of a sub-basin and do not change the prevailing water allocation policies. A plan may recommend how competing uses may be more effectively managed or how conservation might be improved, but these recommendations operate in the margins of the water allocation rules specified in the Water Act and the SSRB Water Management Plan. This is not to suggest that the watershed management plans are unimportant, but they are not as consequential as the legislative and regulatory water management framework of the SSRB.

The government of Alberta launched its Our Water, Our Future initiative in February 2013. Our Water, Our Future is described as a "renewed dialogue" with Albertans about how water is used and governed in the province (Government of Alberta, "Our Water, Our Future: A Conversation" 2014, 10). It involved extensive stakeholder and public consultations on water issues, including special consultations with Metis and First Nations, and involved a variety of means for conveying public input, including various forms of social media. Along with the longstanding issues related to scarcity in the south, newer water issues such as pollution from the northern oil sands and hydraulic fracturing were also prominent in the conversation (ibid.).

In November 2014, the government released a summary of the consultations and an action plan with commitments to address issues of concern that emerged from the consultations. With respect to the SSRB, the government committed to studying the potential for further water storage, to undertaking a pilot project for using existing storage capacity on the Bow River more effectively, and to studying the implications of climate change for water management in the basin (Government of Alberta, "Our Water, Our Future: A Plan" 2014). The emphasis on water storage is interesting and was likely prompted by the devastating floods of 2013. Dams are a potential means of flood control, and this may be sparking renewed interest in them. The same could said of climate change, which will likely result in more erratic precipitation patterns, the effects of which could be smoothed out with more dams and storage capacity. A report on storage construction opportunities in the SSRB was completed by Alberta Agriculture, in cooperation with the Irrigation Council, in the summer of 2014, but no actions were taken. Nevertheless, these recent developments underline the fact that the shift to softer water policy in southern Alberta was never inevitable, and that a shift back towards harder water policy could very well still occur.

As this study has shown, the crucial factors determining the future of Alberta water policy will be the advocacy coalitions prevailing in the water policy subsystem, the balance of power between them, and historical contingencies that open paths for policy reform. The Aggie-Green dynamic has remained quite prominent in the subsystem since the conclusion of the SSRB Water Management Plan, as evidenced by the struggle over amendment transfers, and by the membership of the Watershed Planning and Advisory Councils. The Aggies continue to hold the balance of hard power, though this advantage seems to be slowly bleeding away as the province continues to urbanize and its economic future seems tied to the northern oil sands and the economic importance of agriculture continues to shrink.

In recent years, the water policy subsystem has also been buffeted by major political changes. In the 2012 election, the irrigation belt, with the exception of the two Lethbridge ridings, deserted the PCs en masse in favour of the upstart Wildrose Party. This was the first time since 1975 that the irrigation belt did not go Conservative, and the immediate effect was to reduce the irrigators' influence inside the governing caucus and Cabinet. An even bigger change took place in 2015 with the election of a majority New Democratic Party (NDP) government, a Wildrose opposition (with most of the irrigation belt seats), and the reduction of the PCs to third-party status. The NDP is probably the most sympathetic to the Greens of all Alberta's mainstream political parties, so it is likely that the Greens will have more influence in the governing caucus and Cabinet than they have ever had before. This could constitute an external shock to the water policy subsystem, shifting the balance of hard power, and opening a path for major water policy change. It is too early to tell if this is the case, but two factors make this unlikely. First, the Aggies still effectively control a block of legislative seats while the Greens do not, so the NDP may be sympathetic to the Greens, but the electoral incentives still favour the Aggies. Second, water policy change does not seem to be high on the NDP's policy agenda and the new government is preoccupied with so many other policy issues that it is unlikely to shift its attention to water. So the balance of hard power still seems tilted in the Aggies' favour, and the external shock of the 2015 election is likely to pass without any major water policy changes.

Looking forward, one crucial aspect of the water policy subsystem is the role of the cities. The data in this study showed that, up to 2005, the growing urban and suburban areas played a minor role in water policy development and had not yet formed an advocacy coalition. Since 2005, the cities and suburbs of southern Alberta have continued to grow at some of the fastest rates in Canada, and many of them have been active in the water market, purchasing water licences to support their growth. In fact, many of the water controversies in southern Alberta over the past decade have involved growing suburban areas, like Okotoks outside of Calgary, who have desperately needed water and have had difficulty getting it. Okotoks has outgrown its water licence (which supports about 30,000 people) in recent years and has approached the City of Calgary about the possibility of importing water through a new pipeline. However, the pipeline would cost around $200 million to construct, and the province has been reluctant to foot the bill (CBC News 2014). Other suburban cities around Calgary and Lethbridge are in similar situations, and it would not be surprising if they mobilized to form an urban-suburban advocacy coalition to advance their interests, if they have not done so already.

The emergence of an urban-suburban coalition could completely realign the balance of power in the water policy subsystem in both predictable and unpredictable ways. An urban-suburban coalition would have electoral and legislative hard power that would rival the Aggies', potentially resulting in their displacement as the subsystem's dominant coalition. It may also create a subsystem of shifting coalition alliances in which two coalitions make common cause in pursuit of shared policy goals. It is conceivable, for instance, that the cities might make common cause with the Aggies in pursuit of further dam construction, or that the cities might join with the Greens in demanding more water for environmental flows. All of this is as yet quite speculative, but it does seem likely that, at some point, the urban and suburban areas will become more assertive in the governance of southern Alberta's water, just as cities in other water scarce places have already done.

Testing the ACF

This study was not only an investigation of southern Alberta's shift to softer water policy, it was also a test of the ACF and its applicability to policy development in Canada. As outlined in chapter 2, the ACF has been applied only a handful of times to Canadian cases, and rarely with the sort of methodological rigour used here. Accordingly, this book has provided one of the stiffest tests yet of the ACF's efficacy for studying Canadian policy development. It has also provided a test – the latest in a long line of tests – of the ACF's portability outside of its originating jurisdiction, the United States. Portability has long been a concern of ACF scholars, who have repeatedly revised the framework in light of empirical findings from applications in various parts of the world in an effort to make the ACF as generalizable as possible (Jenkins-Smith et al. 2014, 184–8).

In terms of the ACF's applicability to the Canadian context, the framework holds up very well: the ACF's basic conceptualization of how policy development occurs has been borne out in the southern Alberta water policy case. There is very strong evidence of a southern Alberta water policy subsystem and of the presence of advocacy coalitions in the subsystem, each espousing a different set of policy core beliefs. There is also overwhelming evidence of the coalitions vying with each other to influence policy decision-makers in an effort to have their beliefs reflected in policy. There is also clear evidence of the coalitions using various and different power resources in their attempts to influence policy decision-makers, and of the coalitions trying to outmanoeuvre each other in these attempts. In short, the ACF conception of policy development is found to be quite applicable to Canada, and the ACF may provide a promising vein for future research on Canadian policy development in any number of policy fields.

This study has also yielded some useful insights for the ongoing theoretical development of the ACF in general, and for the line of theoretical development concerned with explaining major policy change in particular. As described in chapter 2, the ACF is an analytical framework that has inspired three distinct yet interrelated lines of theoretical work: a line of theory on the formation and maintenance of advocacy coalitions, a line of theory on policy-oriented learning, and a line of theory on policy change (Jenkins-Smith et al. 2014, 195–204). The last line has been the primary focus of this volume, though the relevance of the other lines is obvious, given the importance of advocacy coalitions and policy-oriented learning in the policy changes recounted here. The theoretical work on policy change has focused on identifying the factors that are necessary for a major policy change to occur, and much of this work has been oriented around testing and refining two central hypotheses, versions of which have been empirically investigated in this book.

The first ACF policy change hypothesis tries to account for the historical contingencies that can reshape policy subsystems and create the necessary conditions for major policy change. In essence, this hypothesis identifies four paths to major policy change and posits that one or more of these paths is necessary for major policy change to occur: "Significant perturbations external to the subsystem, a significant perturbation internal to the subsystem, policy-oriented learning, negotiated agreement, or some combination thereof are necessary, but not sufficient, sources of change in the policy core attributes of a governmental program" (Jenkins-Smith et al. 2014, 203).

Since, in the ACF literature on policy change, "there has been strong support for the first policy change hypothesis," this hypothesis was adopted and tested unmodified in this study (Jenkins-Smith et al. 2014, 203). Specifically, the first policy change hypothesis was investigated in the process tracing analysis of chapters 6, 7, and 8.

Overall, the findings of this study are consistent with the ACF literature in that the first policy change hypothesis was largely supported. All of the major water policy changes in southern Alberta, whether in the Water Act or the SSRB Water Management Plan, found their origin in at least one of the four paths identified as necessary for major policy change to occur. So the basic causal claim of the first hypothesis is supported. Yet the detailed examination of policy change processes afforded by a book-length study also revealed some limitations of the policy change paths as explanations of major policy change, where further theoretical development might take place.

Each policy change path captures a distinctive process that, once initiated, can lead to major policy change. The processes are qualitatively different from each other and they have some empirical validity, as the findings in this and other ACF policy change studies attest. Yet the paths are also caricatures without clearly defined causal chains, and this becomes problematic when one tries to apply them to complex empirical realities, because the paths offer only a general guide to how policy change occurs. The more paths that are empirically present, the more problematic this becomes, as the paths intersect and interact in unspecified ways and the causal chains become longer and more complex. In other words, the paths capture something essential about policy change processes, but they are a bit vague about what they have captured.

Take, for example, the development of the SSRB Water Management Plan, as described in chapter 8. This policy change process started with an internal shock (implementation of the Water Act), was buffeted by an external shock (the 2000–2 drought), was strongly influenced by policy-oriented learning in both coalitions (about water trading, conservation holdbacks, and licensing moratoria), and ended in negotiated agreement between the coalitions. All four policy change paths were present and played a causal role, telling us something important about how this policy was developed. But the causal chain resulting from the application of the paths is blurry, at best. The relative importance of the four paths in this instance is not known, nor is it clear whether or how the paths intersected, or whether the paths were mutually reinforcing or countervailing. Clearly specifying the causal chains at work in the policy change paths would go a long way in answering these questions.

A related limitation of the policy change paths is the lack of specification as to why some paths open but no major policy change occurs. This was a problem confronted in chapter 6, when the PCs first took office in Alberta, and a number of reasons were identified as to why this external shock produced no major water policy changes. Weible (2006) also encountered this issue in his study of California's marine protected areas and similarly speculated on some of the reasons behind it, based on findings from his case. However, there is no general explanation as to why policy paths fizzle out, and the main reason is the under-specification of the policy change paths. That is, somewhere between the start of a path and a prospective policy change, change fizzles out because it is blocked by an intervening factor, factors that are still largely unspecified. So specifying the causal chains in the policy change paths would not only provide a more fulsome explanation of policy change, it would also help to explain policy stability.

Specifying these causal chains is beyond the scope of what can be accomplished here, but, as a starting point, it might be worth distinguishing between "trigger" variables (those that initiate a policy change process) and "tipping" variables (those that tip a subsystem towards policy change). Some trigger variables are suggested by the paths as currently formulated and would include shocks (internal and external), learning events, and hurting stalemates, among others. Tipping variables are causally closer to advocacy coalitions and would include variables such as coalition beliefs, coalition resources, and probably others. There are also some variables, such as policy brokers and coalition opportunity structures, that were found to be important variables shaping policy change in this study and may be included as either trigger or tipping variables, though it is not yet clear into which category they should fall. Distinguishing between trigger and tipping variables would allow for more clearly specified causal chains, and it would create a hypothesis for why some policy paths fizzle out: that is, a trigger variable was moved, but a tipping variable was not. A formulation of this sort might also allow for greater understanding of how policy paths interact and what causal chain(s) is characteristic of each path, and it may facilitate the identification of new policy paths, if some exist. All of this is quite speculative, but it does represent a potential research frontier in the development of ACF explanations of policy change.

The second ACF policy change hypothesis tries to link advocacy coalitions with major policy change. The original hypothesis posits that a change in the dominant coalition, or imposition by a hierarchically superior government, is necessary for major policy change, but the ACF policy change literature, and the ACF water governance subliterature, show only partial support for this hypothesis. Thus, in chapter 2, it was argued that the second policy change hypothesis was in need of revision. Using the distinction between hard and soft power, a revised hypothesis was developed, which stated, "A shift in the balance of coalition hard power, a shift in the balance of coalition soft power, imposition by a hierarchically superior jurisdiction, or some combination thereof, are necessary, but not sufficient, sources of change in the policy core attributes of a governmental program."

The revised hypothesis retains the elements of the original hypothesis that are supported by the literature (that shifts in the balance of hard power and imposition by a hierarchically superior government can cause policy change) and adds a new element (shifts in the balance of soft power) that should expand the explanatory power of the hypothesis. Since imposition by a hierarchically superior government was ruled out as a cause of Alberta water policy change in chapter 3, and a shift in hard power was ruled out as a causal factor in chapter 5, the explanatory factor tested in this study was a shift in soft power.

Overall, the findings in this study support the notion that shifts in soft power can be a cause of major policy change. The survey of coalition resources in chapter 5 makes a good case that a shift in soft power towards the Greens began in the late 1980s and early 1990s, and the process tracing analysis in chapters 6, 7, and 8 clearly shows how the Greens used this soft power to push for water policy reforms. It is equally clear that the balance of hard power remained decisively in favour of the Aggies and that the federal government was little involved in water reform, so that the shift in soft power was the only causal factor at work from the revised second policy change hypothesis. This suggests that shifts in soft power are worthy of further investigation by ACF policy change scholars, though much more empirical research and confirmation is needed before the revised hypothesis can be accepted as a building block in an ACF theory of policy change.

Moreover, the key concepts in the revised hypothesis are not without problems, particularly in operationalization. The approach used in this study was to operationalize the balance of hard power and soft power with reference to the six coalition resources currently identified in ACF scholarship. Though this proved useful for identifying discrete power sources, it was less useful in determining the overall balances of power between the coalitions. This was not problematic in determining the balance of hard power, since the balance was so obviously lopsided in favour of the Aggies. However, it was more problematic in determining the balance of soft power, since both coalitions had some soft power resources, and there was no rigorous, objective guide for determining the relative power provided by these resources. This was resolved by examining and comparing the authoritative, rhetorical, and popular support that each coalition could muster in support of its respective policy positions – which showed a balance of soft power favouring the Greens – but a more rigorous procedure for operationalizing and measuring the balances of hard and soft power would be a welcome addition to this line of research.

Assuming it proves valid beyond this initial case, the revised second policy change hypothesis may also offer new avenues of theoretical development for ACF policy change scholars. It may be possible, for instance, to link different subsystem power configurations with different types of policy change. During Alberta's shift to softer water policy, the subsystem power configuration was characterized by one coalition holding the balance of hard power and another coalition holding the balance of soft power. As observed several times throughout the book, this resulted in policy change characterized by institutional layering, as the Greens' favoured soft reforms were added on top of the Aggies' preserved water licences and water management infrastructure. If the subsystem power configuration had been different – say, the Greens had held the balance of both hard power and soft power – then it is likely that the type of policy change would also have been different. This claim is hypothetical and speculative, but it does speak to the increased explanatory power of the revised hypothesis versus the original one.

The revised second policy change hypothesis may also offer the opportunity for increased synthesis between the two ACF policy change hypotheses that, so far, have been treated in relative isolation. As intimated above, shifts in hard and soft power may be key "tipping" variables that have yet to be specified in the policy change paths contained in the first policy change hypothesis. It is easy to envision, for instance, that an external or internal shock could shift the balance of hard or soft power in a policy subsystem, thereby facilitating major policy change. This not only integrates the two hypotheses but also helps to address some of the shortcomings in the first policy change hypothesis, noted above. This is just one example of potential synthesis of the two policy change hypotheses, still leaving much to do. But eventually ACF scholars will have to work out this synthesis if they hope to move beyond loosely connected hypotheses towards a formal ACF theory of policy development.

Overall, this study finds empirical support for the first ACF policy change hypothesis and for a revised second policy change hypothesis, but also highlights some limitations of the current ACF theoretical work on policy change. Whether it is clarifying the causal chains embodied in the policy change paths, identifying the factors that cause a policy change path to fizzle out, testing and refining the revised second policy change hypothesis, or working for some kind of synthesis between the policy change hypotheses, there are plenty of research frontiers left to explore for ambitious ACF scholars of policy change.

Shedding Light on a Wicked Problem

In chapter 1, water governance in the modern world was characterized as a wicked problem. Wicked problems are wicked because they are not readily solvable. They usually involve a complex web of interdependent social and environmental factors that are not easily reducible and are constantly changing. This produces a high degree of scientific uncertainty and invites actors to try to make sense of the problem based on their own values and world views. When multiple world views are present, as is typical of a wicked problem, the result is competing problem definitions, so that stakeholders cannot even agree on the problems to be addressed, much less the solutions to be adopted. In support of their positions, stakeholders often point to conflicting sources of evidence and advocate a wide range of policy solutions, none of which can be considered as objectively or scientifically optimal as the result of conflicting and plausible policy goals (Balint et al. 2011).

Water governance, particularly at large scales, is wicked because it involves a high degree of scientific uncertainty and a large variety of actors who to try to make sense of it based on diverse and conflicting world views. The Aggies and Greens, for instance, approached water governance in southern Alberta with starkly different values and objectives, and advocated for conflicting policy programs based on these values and objectives. The prevailing scientific knowledge was used by these actors to buttress these respective policy positions as much as they could, but it did not provide objective or technical grounds for choosing between them. The Aggies' objectives of development and prosperity and the Greens' objectives of protection and preservation were both laudable and normative, so that the contest between them was ultimately a political one.

This underscores an important point about wicked problems that is vividly illustrated in this volume: almost everything in a wicked problem is political, from problem definitions to policy solutions. Though an increasing number of scholars have come to characterize water governance as a wicked problem, very few have tried to come to terms with the wicked politics of water governance in a way that makes these politics more intelligible (Reed and Kasprzyk 2009; Weber, Memon, and Painter 2011; Gupta et al. 2013). However, the ACF was designed explicitly for analysing policy development in wicked problem situations, and the ACF analysis in this volume suggests that there are some regularities in the politics of water governance that can shed some light on the dynamics of this wicked problem (Sabatier and Weible, "Advocacy Coalition Framework" 2007, 189).

The first dynamic to note, and one that is supported by other ACF analyses of water governance, is the tendency for actors with similar world views to coalesce into advocacy coalitions. This may seem like a facile observation, but it is not. This tendency towards advocacy coalitions indicates that there is some social order in wicked problems that can otherwise seem disorderly, confusing, and even chaotic. The number and types of advocacy coalitions vary in different contexts, but the tendency towards advocacy coalitions is quite clear. This suggests that observers and participants in water governance situations would do well to step back and take stock of the advocacy coalitions in their midst if they want to understand the basic political dynamics of the wicked problem in which they are embroiled.

Another dynamic in wicked problems is the importance of political power, in all its forms. This observation may also seem a bit facile, but the essentiality of political power in water governance is too often overlooked, particularly in common pool studies focused on collective action and institutions of collective action (Ostrom 1990). The southern Alberta case shows how water follows political power: when the Aggies had a monopoly on political power, they accumulated southern Alberta's water for decades; when the Greens emerged and some political power shifted to them, the Aggies' accumulation of water stopped and even slightly reversed. It seems to take some time for water to follow political power – it took a number of years for Green soft power to have some policy impact – so that there is a lagged relationship between these two variables. But the importance of political power seems clear from this study, and appears to be a theme in other studies of water governance that seriously analyse the politics of water and water allocation (Worster 1985; Ingram 1990).

Political power is important, but water governance is more than just a contest of raw political power; it also involves actors trying to make collective sense of the problems besetting them and crafting acceptable solutions to these problems. The challenges here are formidable, but some factors stand out in the southern Alberta case as particularly important to these processes.

One is the use of inter-coalition consultations and deliberations, away from the partisan and electoral politics of the legislative arena. These processes were shown to be more soft-power-friendly, encouraging discussions of the public good and the evaluation of policy options on these terms. This finding supports the arguments of many wicked problem scholars who typically advocate such processes for the governance of all sorts of wicked problems (Balint et al. 2011). It also meshes well with the ACF scholarship on policy-oriented learning, which has found that policy forums that bring together the leadership of opposing advocacy coalitions in a professional environment are a key factor in facilitating cross-coalition learning (Jenkins-Smith et al. 2014, 200).

Another important factor was the presence of a policy broker. Alberta Environment increasingly played this role through the development of the Water Act and the SSRB Water Management Plan, and, in doing so, it helped the coalitions reach common understandings and negotiate acceptable policy compromises. Policy brokers have received limited attention in the ACF literature and almost no attention in the wicked problem literature, but clearly seem to warrant further study.

Ultimately, the findings from this study do not "solve" the wicked problem of water governance. There are no optimal policy solutions to wicked problems, but there are policies that are better or worse for all involved. The findings here shed some light on how policies are adopted in wicked problem situations, findings that can be built on by others to help actors avoid the worse policies and make progress towards the better ones.

Appendix: Measuring Policy Core Beliefs Using Qualitative Content Analysis

For readers interested in the social science behind this book, this appendix provides a detailed account of the content analysis methodology used in chapter 4. The purpose of the content analysis was to measure policy elites' policy core beliefs at different points in the thirty-year period of study. Policy core beliefs are a key explanatory variable in the ACF. Policy core beliefs are the glue holding advocacy coalitions together, and advocacy coalitions endlessly strive to see their policy core beliefs enshrined in government policy. Accordingly, getting a clear and accurate picture of the prevailing policy core beliefs, and how they have changed over time, is crucial to any ACF analysis.

ACF analysts must take considerable care in generating data about prevailing policy core beliefs. In their review of the ACF literature, Weible et al. suggest that too many ACF studies have "appeared to rely on unsystematic collection and analysis of existing documents and reports" (Weible, Sabatier, and McQueen 2009, 126). Instead, "the goal should be intersubjectively reliable methods that serve as the cornerstone for making research transparent and provide a basis for collective learning" (135). This was the goal of this study, and it was pursued through a systematic content analysis of stakeholder submissions at three points in time: 1978, 1995, and 2002/5. Why and how I undertook this analysis is explained in detail below.

Content analysis is a commonly used data analysis technique in the ACF literature precisely because it is a systematic approach for discerning meanings in qualitative texts. Any well-constructed content analysis is superior to more casual data interpretations because it applies a standardized coding frame to large volumes of qualitative data, helping researchers to sort and summarize textual material in a relatively objective manner. In this way, content analysis helps to minimize the subjective biases of researchers interpreting qualitative texts, providing a more valid and reliable understanding of patterns and themes within qualitative data (Schreier 2012).

The first challenge was to locate raw data sources from which the policy elites' policy core beliefs could be measured. The best ACF scholarship has typically relied on data from one or more of three sources: elite surveys (Weible and Sabatier 2009), elite interviews (Ingold 2011), or transcripts of legislative debates and public hearings (Nohrstedt 2008; Pierce 2011). Because some of the policy processes being analysed in this study occurred almost forty years ago, locating stakeholders to administer a survey or conduct interviews was unviable. I was forced, instead, to rely on archival materials and was quite fortunate that an abundance of primary source material was available. Public consultations on water issues were held in southern Alberta in 1978, 1984, 1995, and 2002/5. Transcripts of the 1978, 1984, and 1995 hearings/submissions were available through the University of Alberta library, and copies of the 2002/5 submissions were obtained through a Government of Alberta Freedom of Information request. Through all of the analysis, the 2002 and 2005 submissions were treated as a single round of consultations, because they were two phases in one round of water planning, the development of the SSRB Water Management Plan.

These data sources yielded thousands of pages of raw qualitative data – far more than could be feasibly analysed – creating a formidable sampling challenge. The raw data set comprised stakeholder testimony and submissions from four different years. The sampling unit was defined as an "actor-year": one actor's testimony in one year. The goal of sampling was to delineate the set of actor-years that would be most useful in answering the research question (Krippendorf 2013, 99–100). As is common practice in content analysis, the data were sampled on the basis of relevance rather than representativeness. That is, rather than trying to generate a representative sample, a sample was generated on the basis of its relevance to the research question, in this case, the data most relevant to revealing the policy core beliefs of key elites in the policy subsystem (120–1). Relevance was determined on the basis of four factors:

1. Recurring actor participation. As Norhstedt points out, "The ACF emphasizes frequent participants in policymaking, who are differentiated from one-shot participants that only participate on single

occasions" (Nohrstedt 2011, 468). In their studies, Zafonte and Sabatier (2004) and Nohrstedt (2011) define frequent participants as those actors participating in official policy development/consultation processes (e.g., public hearings) on at least three occasions. However, they were analysing policy subsystems with more frequently occurring processes, with more opportunities for actors to participate. For this study, this approach was followed, but it defined frequently occurring participants as those actors making submissions in at least two of the 1978, 1984, 1995, and 2002/5 public hearings in the southern Alberta water policy subsystem. Three sub-issues also came up when defining frequently occurring participants:

a. Many ACF analysts have confined their analyses to examining organizations, as opposed to individuals, but both organizations and individuals were included in the sample for this study, to ensure that some of the early Greens who may have participated in policy development as individuals prior to the formation of Green organizations were not excluded from the analysis.

b. Special care was taken to ensure that those organizations that underwent a name change during this period – such as the Alberta Cattle Commission renaming itself the Alberta Beef Producers – were not excluded from the sample on that basis.

c. The 2002/5 round of policy development was different from the others in that it relied more heavily on co-management bodies known as basin advisory committees. These basin advisory committees allowed a wide range of policy elites to have direct involvement from start to (almost) finish in developing the SSRB Water Management Plan, ultimately making recommendations on what should be included in the final plan. The process also featured open public consultations, but, with so many of the elites already involved on the basin advisory committees, fewer felt the need to make public submissions. So the submissions from this round under-represent the elites who were involved in the process. To compensate for this, participants from this round were defined as anyone who made a public submission and/or sat on a basin advisory committee.

Once these issues were settled, lists of participants (both organizations and individuals) in the four rounds of hearings were created and cross-referenced to determine who had participated at least twice. The submissions of the recurring actors were included in the sample, and all others were excluded.

2. Removal of redundancies. Some organizations had multiple actors testify on their behalf in a given round of public hearings. The Alberta Irrigation Projects Association, for example, had eight actors testify on its behalf during the 1978 hearings. Redundant testimony such as this was removed from the sample in the interest of efficiency. Where there were redundancies, the submissions were examined to determine which was the "official" submission of the organization, and that one was retained. If an organization's "official" submission could not be determined, the submission of the highest-ranking person within the organization was retained.

3. Key points in time. Although four rounds of public consultations were undertaken in the policy subsystem, it was necessary only to analyse three rounds of public consultations in order to gain an accurate picture of the evolving policy core beliefs throughout the period of study. To this end, the 1984 hearings were excluded from the sample because they were only six years removed from the 1978 hearings, and because there was no major water policy change in the 1980s to make examination of the 1984 hearings worthwhile. Excluding the 1984 hearings left a sample that included three crucial points in time: 1978, when hard policy was still prevalent; 1995, during the development of the Water Act, when softer policy began to emerge; and, 2002/5, during the development of the SSRB Water Management Plan, when softer policy instruments were put into action. These three points in time were sufficiently spread out and sufficiently spaced to provide three good reference points to chart any changes in policy core beliefs during the period of study. Though the 1984 public hearings were excluded from the sample, it should be reiterated that the list of participants from the 1984 hearings was still used to help determine frequent participants in the subsystem, in the manner described above.

4. Focus on submissions. There was something of an equivalency challenge in the data in that the 1978 data were in the form of testimony transcripts, while the 1995 and 2002/5 data were in the form of written submissions. Fortunately, this problem was easily resolved. The 1978 testimony followed a pattern in which actors first read a statement into the record, followed by a question-and-answer period between the testifying actor and the committee conducting the hearings. Simply dropping the question-and-answer period from each actor's testimony left submissions that were similar to those from 1995 and 2002/5. Thus, in all three rounds, the submissions provided a means for actors to state their policy beliefs unilaterally, without prompting or direction by interrogators.

The only real (but relatively inconsequential) difference was that the 1978 submissions were oral while the 1995 and 2002/5 submissions were written.

On the basis of these four parameters, a manageable and relevant sample of elite submissions was created. This constituted the raw qualitative data that would be content analysed to determine the prevailing policy core beliefs of key actors in the Alberta water policy subsystem.

A crucial decision in designing the content analysis was whether to undertake a qualitative or quantitative analysis of the sampled texts. Ultimately, a qualitative approach was chosen because of a problem with a key assumption of quantitative content analysis: frequency as an indicator of intensity. The typical approach in quantitative content analysis is to undertake counts of words/phrases/ideas in texts, develop frequency tables, and then use frequency as a proxy for the intensity with which these beliefs are held. In this way, nominal level qualitative data can be transformed into interval or ratio level quantitative data that can be statistically analysed. Once I familiarized myself with the data I was to analyse, I found that the fundamental assumption of quantitative content analysis did not hold; that frequency was not a good proxy for intensity. In the submissions analysed, the frequency of stated beliefs had more to do with the internal logical structure of the submissions' arguments and the authors' rhetorical styles. Some submissions were very brief and terse, while other submissions were lengthy and verbose. Some submissions responded directly to a report or a draft bill, while others did not. In this situation, it made little sense to claim that actor X supported dam construction twice as much as water markets because the former was mentioned twice and the latter once. It made even less sense to claim that actor X supported dam construction twice as much as actor Y because the former mentioned it twice and the latter mentioned it once. All of these differences could easily be accounted for by rhetorical structure and style. Accordingly, the assumption that frequency equalled intensity did not hold up, so I opted for a qualitative study that analysed the presence and absence of policy beliefs rather than the intensity with which they were held.

Another important decision in designing the content analysis was to select a coding unit. Coding units may be defined physically, such as when words, sentences, or paragraphs are used as coding units, or conceptually, such as when ideas or themes are used as coding units (Krippendorf 2013, 100–1). Following the lead of Leifeld (2013), this study used a conceptual coding unit, defined as a "policy recommendation." Most actors made submissions to influence government policy and, in that effort, they made policy recommendations. These recommendations were based on how elites perceived policy problems and what solutions they would like to see ensconced as public policy. Accordingly, it was felt that policy recommendations were a close proxy for policy core beliefs.

In its internal structure, any policy recommendation has two key elements: a concept and an expression of disposition (e.g., support or opposition) for that concept. Concepts refer to a specific water policy action or inaction. Broad and general concepts – such as "sustainability," "economic growth," "a healthy environment" – were excluded from the analysis, because they have manifold meanings and do not point to specific policy actions. Some irrigators, for example, claimed to support a healthy environment but then made specific policy recommendations that were contrary to this assertion. Similarly, some environmentalists claimed to support the growth of rural economies but then made specific policy recommendations that made this nearly impossible. The aim was to cut through this sort of rhetoric and focus on what actors truly believed should be in water policy. Since broad and general concepts were unreliable indicators, the focus was on specific and actionable concepts instead. Concept disposition refers to whether an actor is supportive, unsupportive, or neutral about a concept. Most actors mention a concept only in order to express their support or opposition to it. Accordingly, disposition is a key part of determining policy core beliefs.

In addition to coding units, most content analysts also define context units, which are (larger) segments of text to be taken into consideration when trying to code individual coding units (Krippendorf 2013, 101–2). This analysis was somewhat unusual, in the sense that the defined context unit was located outside the texts being analysed. In all three rounds of public submissions analysed, the actors making submissions were responding to specific policy documents: a report on water development in the Oldman sub-basin in 1978, a piece of draft water legislation in 1995, and a draft water planning document for the SSRB in 2002/5. Accordingly, these specific policy documents were crucial for understanding the public submissions in their proper context, so they were defined as the context unit for the content analysis. Coders were provided copies of these documents, read them prior to coding, and could refer to them during coding.

One of the biggest research design challenges for this analysis was the creation of a coding frame. A coding frame is a hierarchical framework of categories used as a decision tree when coding each segment of text in a sample. The coding frame used in the analysis was based largely on the internal logical structure of the coding unit. Policy recommendations (the coding unit) involve actors establishing a concept and communicating their disposition towards it. This suggested a two-level coding frame: one for distinguishing concept types and another for determining an actor's support, opposition, or neutrality towards the concept. The general structure of the coding frame is shown in figure A.1.

As shown, the first level of the coding frame distinguished between different types of concepts. In keeping with the aim to cut through political rhetoric, the focus was on specific and actionable concepts rather than broad and general concepts. Since there was no a priori way to develop a complete list of concepts, the list was developed through immersion in the data. The entire sample was read once with an eye to determining concepts: each time a potential new concept was encountered, it was noted in a list. The concepts in this list were then subjected to subsumption, eliminating duplications, and folding substantively similar concepts into each other, as much as possible (Schreier 2012, 115–17). The concepts were also analysed and revised to achieve mutual exclusiveness. In the end, the first level of the coding frame identified 102 distinct policy concepts, allowing for quite precise coding.

The second level of the coding frame distinguished between different dispositions towards an identified concept. Regardless of the concept identified, there were four coding options in terms of disposition: supportive, unsupportive, neutral, and unable to determine disposition. Coders encountered various levels of disposition – from passionate to begrudging – but all levels of disposition were coded as simply "supportive" or "unsupportive." Coders also encountered expressions of conditional disposition and, unless the conditions were judged as onerous or impossible, these were also coded as simply "supportive" or "unsupportive." "Neutral" codes applied to instances when an actor stated no position on a concept or explicitly stated neutrality on a concept. The "unable to determine disposition" code applied when actors expressed contradictory dispositions (e.g., support, then opposition) for a concept, or did not describe their disposition clearly.

Data analysis began with segmentation. Segmentation involves identifying each occurrence of the coding unit in the raw data (Schreier

Figure A.1 Coding Frame Structure

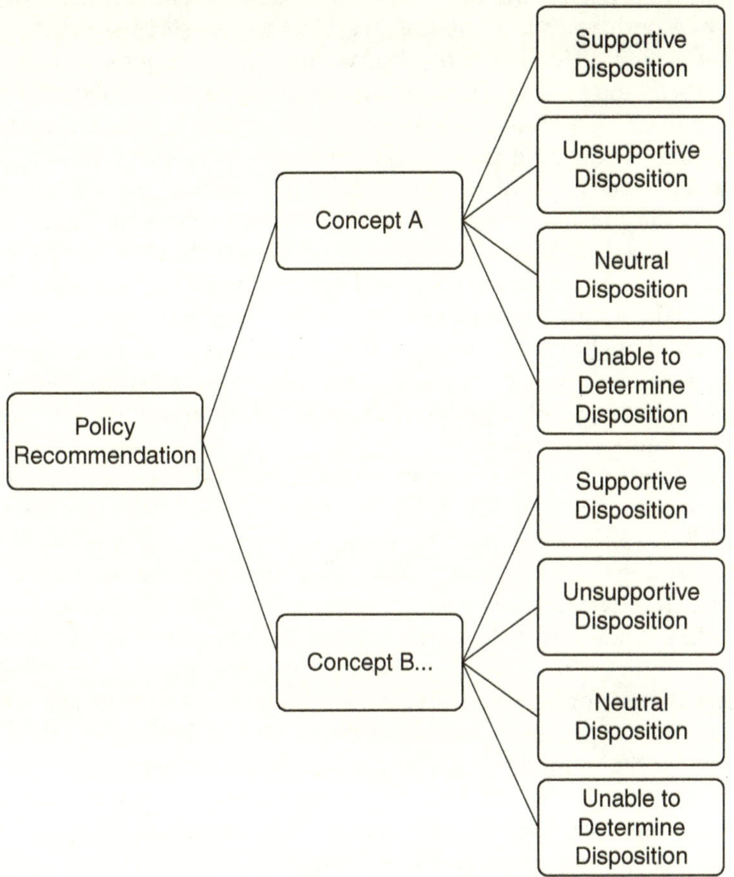

2012, 127). This meant reading all submissions in the sample, marking each policy recommendation in square brackets, and labelling each policy recommendation with a unique identifier. Given the considerable contextual and technical knowledge required to readily and properly identify policy recommendations, I undertook this task myself. This was a labour-intensive process, but it produced 974 policy recommendations that could be readily coded using the established coding frame.

The coding itself involved two coders operating from the same set of coding instructions. I was one coder and the other was a research assistant who underwent training to ensure an aptitude with the coding instructions. Each coder independently coded the entire sample once.

When both rounds of coding were complete, steps were taken to create a single (final) set of observations. Each coder's coding was compared and a coefficient of agreement was calculated (Schreier 2012, 170). The calculations showed 79.6 per cent agreement between the coders, indicating that a high level of inter-coder reliability had been achieved, particularly for a qualitative analysis involving such a large coding frame. All observations showing inter-coder agreement were retained as final observations. All observations showing inter-coder disagreement were discussed and analysed by the coders in an effort to develop a consensus interpretation. Where consensus could be reached, these observations were retained as final observations; where consensus could not be reached, these observations were marked as indeterminable (174). This left a single set of observations with a high level of inter-coder reliability.

This single set of observations provided the data that form the basis for tables 4.1, 4.2, and 4.3, which summarize elites' policy core beliefs in 1978, 1995, and 2002/5, respectively. Three steps were important in transforming the data into the final tables shown in the chapter.

First, the data were cleaned. This involved removing observations coded as "indeterminate," which provided no useful information on actors' policy core beliefs. It also involved the removal of identical observations from each actor-year, since the aim of the study was to determine the presence or absence of policy core beliefs, not their frequency. If the same actor-year expressed contradictory dispositions on the same policy concept (i.e., they both supported and did not support the concept), its position was regarded as indeterminate and these observations were removed from the data.

Second, for ease of presentation, some of the 102 policy concepts in the coding frame that are closely related (and co-varied) were folded into larger concepts. For example, the concepts "expand irrigation acreage," "expand irrigation water use," "expand irrigation infrastructure," "expand irrigation financial support," and "expand irrigation in general" were folded into the concept "expand irrigation." This was done for many other concepts as well, resulting in tables that were easy to interpret but well represented the data.

Third, an analysis of ideological groupings in each year was undertaken. The intent was to undertake a quantitative cluster analysis, but this proved impossible because there were too much missing data in the proximity matrices. These missing data were a by-product of relying on textual data sources in which actors could choose to address some concepts but not others, resulting in a lot of missing values. These missing values are evident in the blank spots in tables 4.1, 4.2, and 4.3. Instead of a cluster analysis, ideological groupings were identified by repeated testing for hypothesized advocacy coalitions: an irrigation coalition, a dryland farming coalition, an environmentalist coalition, a conservation coalition, an urban coalition, an industrial coalition, and so on. As the tables show, the only clear ideological groupings that emerged from this repeated inference testing were the Greens and the Aggies (Alberta Agriculture 2010).

References

Ainuson, Kweku. "An Advocacy Coalition Approach to Water Policy Change in Ghana: A Look at Belief Systems and Policy Oriented Learning." *Journal of African Studies and Development* 1, no. 2 (2009): 16–27.

Alberta Agriculture. *Alberta Irrigation Information: Facts and Figures for the Year 2009*. Edmonton: Government of Alberta, 2010.

– *Irrigation in Alberta*. Edmonton: Government of Alberta, 2000.

Alberta Agriculture and Rural Development. *Irrigation in Alberta: Statistical Overview – 2012*. 13 August 2013. http://www1.agric.gov.ab.ca/$department/deptdocs.nsf/all/irr13223.

Alberta Agriculture, Food and Rural Development. "Irrigation Development in Alberta: Water Use and Regional Development." 2004. http://www.ijc.org/rel/pdf/83_stmary-milk_letter.pdf.

– *Provincially Owned and Operated Reservoirs*. 9 July 2004. Accessed 24 January 2005. http://www1.agric.gov.ab.ca/$deparment/deptdocs.nsf/all/irr8793?opendocument.

Alberta Economic Development Authority. "Sustainable Water Management and Economic Development in Alberta." 2008.

Alberta Environment. "Approved Water Management Plan for the South Saskatchewan River Basin." August 2006.

– "Draft Water Management Plan for the South Saskatchewan River Basin in Alberta." 18 October 2005.

– "Establishment of Bow River Sub-Basin Water Conservation Objectives." 16 January 2007.

– "Establishment of Oldman River Sub-Basin Water Conservation Objectives." 16 January 2007.

– "Establishment of Red Deer River Sub-Basin Water Conservation Objectives." 16 January 2007.

– "Establishment of South Saskatchewan River Sub-Basin Water Conservation Objectives." 16 January 2007.
– "General Policy Principles Governing Water Management in Alberta." 1972.
– "Information Sheet: The Purposes of the Existing Dams in the South Saskatchewan River Basin (Alberta)." n.d. Accessed 13 June 2013. http://www.environment.alberta.ca/documents/InfoSheet_Dams.pdf.
– "Joint Meeting of the Basin Advisory Committees, South Saskatchewan River Basin (SSRB) Water Management Plan: Phase One. Meeting Notes (Calgary) Tuesday, January 22, 2002." 2002.
– *Provincial Reservoir Storage Summary.* 13 June 2013. http://www.environment.alberta.ca/forecasting/reports/Res_Storage.pdf.
– "Public Consultation Plan, Oldman River Basin and South Saskatchewan River Sub-Basin: The Year 2000 Review of Water Management in the South Saskatchewan River Basin." 2000.
– "Southern Region Completed Transfers Summary." May 2010.
– "South Saskatchewan River Basin Water Allocation." 2003.
– *South Saskatchewan River Basin Water Information Portal.* 2013. http://esrd.alberta.ca/water/programs-and-services/south-saskatchewan-river-basin-water-information/default.aspx.
– *South Saskatchewan River Basin Water Management Plan: Phase One Water Allocation Transfers.* Edmonton: Government of Alberta, 2002.
– *Water for Life: Alberta's Strategy for Sustainability.* Edmonton: Alberta Environment, 2003.
– *Water for Life Progress Report: December 1, 2008–March 31, 2011.* Edmonton: Alberta Environment, 2011.
– "Water for Life – Results from Random Telephone Survey: Summary Report." 2002.
– "Water Management in Alberta: Challenges for the Future." Edmonton: Alberta Environment, 1991.
– "Water Resources Management Principles for Alberta." 1978.

Alberta Environment and Alberta Agriculture. "Water Management for Irrigation Use." 1975.

Alberta Environment, Environmental Council of Alberta, and Water Resources Commission. *Water Management in Alberta: Summary of Public Comments, July to December 1991.* Edmonton: Alberta Environment, 1992.

Alberta Environmental Protection. "Discussion Draft of Legislation." 1994.

Alberta Federation of Agriculture. *About AFA.* 2014. http://www.afaonline.ca/about-afa.

Alberta Fish and Game Association. "Organizational Profile." 2014. http://www.afga.org/profile.html.

Alberta Water Resources Commission. "Water Management in the South Saskatchewan River Basin: Report and Recommendations." 1986.

Alberta Wilderness Association. "About Us." 2014. https://albertawilderness.ca/about-us/.

Alberta WPACs. "Alberta Watershed Planning and Advisory Councils." 2015. http://www.albertawpacs.ca/.

Albright, Elizabeth A. "Policy Change and Learning in Response to Extreme Flood Events in Hungary: An Advocacy Coalition Approach." *Policy Studies Journal* 39, no. 3 (2011): 485–511.

Arcand, Alan, and Mario Lefebvre. "Alberta's Rural Communities: Their Economic Contribution to Alberta and Canada." Conference Board of Canada. March 2012. http://www1.agric.gov.ab.ca/$Department/deptdocs.nsf/all/csi14195/$FILE/alberta-rural-communities-report.pdf.

Archer, Keith. "Conflict and Confusion in Drawing Constituency Boundaries." *Canadian Public Policy* 19, no. 2 (1993): 177–93.

Armstrong, Christopher, Matthew Evenden, and H.V. Nelles. *The River Returns: An Environmental History of the Bow.* Montreal and Kingston: McGill-Queen's University Press, 2009.

Babooram, Avani. "Agricultural Water Use in Canada." Statistics Canada. 19 September 2011. http://www.statcan.gc.ca/pub/16-402-x/16-402-x2011001-eng.htm.

Balint, Peter J., Ronald E. Stewart, Anand Desai, and Lawrence C. Walters. *Wicked Environmental Problems: Managing Uncertainty and Conflict.* Washington DC: Island, 2011.

Bankes, Nigel, and Arlene Kwasniak. "The St Mary's Irrigation District Licence Amendment Decision: Irrigation Districts as a Law unto Themselves." *Journal of Environmental Law and Practice* 16, no. 1 (2005): 1–18.

Barnett, J.P., J.C. Adam, and D.P. Lettenmair. "Potential Impacts of a Warming Climate on Water Availability in Snow-Dominated Regions." *Nature* 438 (November 2005): 303–9.

Barnett, Vicki. "River Standards Set before Consultation." *Calgary Herald,* 18 January 1992.

Barrie, Doreen. "Ralph Klein 1992–." In *Alberta Premiers of the Twentieth Century,* ed. Bradford J. Rennie, 255–79. Regina: Canadian Plains Research Center, University of Regina, 2004.

Bennett, Andrew. "Process Tracing: A Bayesian Perspective." In *The Oxford Handbook of Political Methodology,* ed. Janet M. Box-Steffensmeier, Henry E. Brady, and David Collier, 702–21. Oxford: Oxford University Press, 2008.

Bernier, Luc, Keith Brownsey, and Michael Howlett. *Executive Styles in Canadian Government: Cabinet Structures and Leadership Practices in Canadian Government.* Toronto: University of Toronto Press, 2005.

Berzins, W.E., R. Harrison, and P. Watson. "Implementation of the Alberta Water for Life Strategy: Strategic Partnerships in Action." *Water Science & Technology* 53, no. 10 (2006): 255–60.

Bjornlund, Henning, Lorraine Nicol, and K.K. Klein. "Challenges in Implementing Economic Instruments to Manage Irrigation Water on Farms in Southern Alberta." *Agricultural Water Management* 93, no. 3 (2007): 131–41.

– "Changing Tactics for Managing Water in Alberta: The Water for Life Strategy and Challenges in Its Implementation." In *World Environmental and Resources Congress 2006: Examining the Confluence of Environmental and Water Concerns,* 1–10. ASCE, 2006.

Blais, Jules M., David W. Schindler, Derek C.G. Muir, Martin Sharp, David Donald, Melissa Lafreniere, Eric Braekevelt, and William M.J. Strachan. "Melting Glaciers: A Major Source of Persistent Organochlorines to Subalpine Bow Lake in Banff National Park, Canada." *Ambio* 30, no. 7 (2001): 410–15.

Boundary Waters Treaty. 1909. International Joint Commission. http://www.ijc.org/en_/BWT.

Bow River Basin Council. Homepage. 2015. http://www.brbc.ab.ca/.

Bratt, Duane. *Canada, the Provinces, and the Global Nuclear Revival.* Montreal and Kingston: McGill-Queen's University Press, 2012.

Brooks, David B., and Oliver M. Brandes. "Why a Water Soft Path, Why Now and What Then?" *Water Resources Development* 27, no. 2 (June 2011): 315–44.

Brooymans, Hanneke. "Environmentalists Lose 'Prime Mentor' Kostuch." *Calgary Herald,* 24 April 2008.

Brooymans, Hanneke, and Dennis Hryciuk. "Premier Douses Water Plan: Diverting to South a Risky Non-starter." *Edmonton Journal,* 3 January 2002.

Brownsey, Keith. "Enough for Everyone: Policy Fragmentation and Water Institutions in Alberta." In *Canadian Water Politics: Conflicts and Institutions,* ed. Mark Sproule-Jones, Carolyn Johns, and B. Timothy Heinmiller, 133–55. Montreal and Kingston: McGill-Queen's University Press, 2008.

Bukowski, Jeanie. "Spanish Water Policy and the National Hydrological Plan: An Advocacy Coalition Approach to Policy Change." *South European Society and Politics* 12, no. 1 (2007): 39–57.

Byrne, James, Stefan Kienzle, and David Sauchyn. "Prairie Water and Climate Change." In *The New Normal: The Canadian Prairies in a Changing Climate,* ed. David Sauchyn, Harry Diaz, and Surendra Nath Kulshreshtha, 61–79. Regina: Canadian Plains Research Center, 2010.

Cairney, Paul. "Standing on the Shoulders of Giants: How Do We Combine the Insights of Multiple Theories in Public Policy Studies." *Policy Studies Journal* 41, no. 1 (2013): 1–21.

Cairns, Alan C. "The Electoral System and the Party System in Canada, 1921–65." *Canadian Journal of Political Science* 1, no. 1 (1968): 55–80.

Canadian Broadcasting Corporation. "Alberta, Saskatchewan Shelve Plans for Meridian Dam." 11 March 2002. Accessed 22 June 2007. http://www.cbc.ca/canada/story/2002/03/11/meridian_dam020311.html.

Canadian Consulting Engineer. "Funding Flows for Pipeline in Alberta." 30 March 2004. http://www.canadianconsultingengineer.com/engineering/funding-flows-for-pipeline-in-alberta/1000016448/.

Canadian Election Atlas. "Alberta Maps." n.d. http://canadianelectionatlas.blogspot.ca/p/alberta.html.

CBC News. "Water Supply Top Concern for Okotoks as Town Grows." 22 July 2014. http://www.cbc.ca/news/canada/calgary/water-supply-top-concern-for-okotoks-as-town-grows-1.2713951.

Clipperton, Kasey G., Wendell C. Koning, Allan G.H. Locke, John M. Mahoney, and Bob Quazi. "Instream Flow Needs Determinations for the South Saskatchewan River Basin, Alberta, Canada." 2003.

Cradduck, Mervin. *Hearings on Water Management in the Oldman River Basin.* Taber: Environment Council of Alberta, November 1978. 46–62.

CrossIron Mills. "CrossIron Mills Fact Sheet." 2009. Accessed 7 May 2013. www.crossironmills.com/media.

D'Aliesio, Renata. "Tapped Out: Water Woes, Part One." *Calgary Herald*, 1 December 2007.

– "Tapped Out: Water Woes, Part Two." *Calgary Herald*, 1 December 2007.

de Loë, Rob C. "Dam the News: Newspapers and the Oldman River Dam Project in Alberta." *Journal of Environmental Management* 55, no. 4 (1999): 219–37.

– "Stability and Change in Southern Alberta Water Management." PhD diss., University of Waterloo, 1994.

Ellison, Brian A. "The Advocacy Coalition Framework and the Implementation of the Endangered Species Act: A Case Study in Western Water Politics." *Policy Studies Journal* 26, no. 1 (1998): 11–29.

Ellison, Brian A., and Adam J. Newmark. "Building the Reservoir to Nowhere: The Role of Agencies in Advocacy Coalitions." *Policy Studies Journal* 38, no. 4 (2010): 653–78.

Elton, David K., and Arthur M. Goddard. "The Conservative Takeover, 1971–." In *Society and Politics in Alberta*, ed. Carlo Caldarola, 49–72. Toronto: Methuen Publications, 1979.

Environmental Law Centre. "In Response to the Draft Water Conservation and Management Act." In *Water Management Policy and Legislation Review: Public Submissions*, 1–103. Alberta Environmental Protection, February 1995.

Environment Council of Alberta. *Report and Recommendations: Public Hearings on Management of Water Resources within the Oldman River Basin*. Edmonton: Environment Council of Alberta, 1979.

Equus Consulting Group Inc. "South Saskatchewan River Sub-Basin: Consultation on Draft Water Management Plan, October 2005–January 2006." 2006.

– *Water for Life: Summary of Consultation Results*. Edmonton: Government of Alberta, 2002.

Figliuzzi, Sal. "South Saskatchewan River Sub-Basin Contributions to International and Interprovincial Water-Sharing Agreements." 2002.

Gasteyer, Stephen P. "Agricultural Transitions in the Context of Growing Environmental Pressure over Water." *Agriculture and Human Values* 25 (2008): 469–86.

George, Alexander L., and Andrew Bennett. *Case Studies and Theory Development in the Social Sciences*. Cambridge, MA: MIT Press, 2005.

Gerring, John. *Case Study Research: Principles and Practices*. Cambridge, MA: Cambridge University Press, 2007.

– "What Is a Case Study and What Is It Good For?" *American Political Science Review* 98, no. 2 (2004): 341–54.

Gillis, C. "Alberta Water Diversion Plan All Wet: Critics." *National Post*, 29 December 2001.

Glen, Barb. "Southern Alberta Dam Marks 20th Birthday." *Western Producer*, 12 October 2012. http://www.producer.com/2012/10/southern-alberta-dam-marks-20th-birthday%E2%80%A9/.

Glenn, Jack. *Once upon an Oldman: Special Interest Politics and the Oldman River Dam*. Vancouver: UBC Press, 1999.

Godwin, R.B. "The Saskatchewan Nelson Basin Board Study in Retrospect." *Canadian Water Resources Journal* 6, no. 1 (1981): 86–95.

Government of Alberta. "Approved Water Management Plan for the South Saskatchewan River Basin." 2006.

– "The Economy: Economic Results." 2014. http://www.albertacanada.com/business/overview/economic-results.aspx.

– "Our Water, Our Future: A Conversation with Albertans. Summary of Discussions." 2014. http://esrd.alberta.ca/water/water-conversation/documents/WaterFuture-SummaryDiscussions-2014.pdf.

– "Our Water, Our Future: A Plan for Action." 2014. http://esrd.alberta.ca/water/water-conversation/documents/WaterFuture-PlanAction-Nov2014A.pdf.

– "Water for Life: Alberta's Strategy for Sustainability." November 2003.

Government of Canada. *Canada Year Book, 1930*. Ottawa: King's Printer, 1930.

Gray, James H. *Men against the Desert*. Saskatoon: Fifth House Publishers, 1996.

Green Party of Alberta. "Alberta Greens Choose New Leader, New Name!" News release, 1 October 2012. http://greenpartyofalberta.ca/alberta-greens-choose-new-leader/.

Gupta, Joyeeta, Aziza Akhmouch, William Cosgrove, Zachary Hurwitz, Josefina Maestu, and Olcay Ünver. "Policymakers' Reflections on Water Governance Issues." *Ecology and Society* 18, no. 1 (2013): 35.

Halliday, R., and G. Faveri. "The St Mary and Milk Rivers: The 1921 Order Revisited." *Canadian Water Resources Journal* 32 (2007): 75–92.

Han, Heejin, Brendon Swedlow, and Danny Unger. "Policy Advocacy Coalitions as Causes of Policy Change in China? Analyzing Evidence from Contemporary Environmental Politics." *Journal of Comparative Policy Analysis* 16, no. 4 (2014): 313–34.

Heclo, Hugh. *Modern Social Politics in Britain and Sweden*. New Haven, CT: Yale University Press, 1974.

Heinmiller, B. Timothy. "Advocacy Coalitions and the Alberta Water Act." *Canadian Journal of Political Science* 46, no. 3 (2013): 525–47.

– "Multilevel Governance and the Politics of Environmental Water Recoveries." In *Multilevel Environmental Governance: Managing Water and Climate Change in Europe and North America*, ed. Inger Weibust and James Meadowcroft, 58–79. Cheltenham: Edward Elgar, 2014.

Heritage Community Foundation. "Political Institutions and Process: Representative Democracy in Action." n.d. http://wayback.archive-it.org/2217/20101208160520/http://www.abheritage.ca/abpolitics/process/election_results.html.

Homer-Dixon, Thomas. *Environment, Scarcity and Violence*. Princeton: Princeton University Press, 1999.

Horbulyk, Theodore M. "Liquid Gold? Water Markets in Canada." In *Eau Canada: The Future of Canada's Water*, ed. Karen Bakker, 205–18. Vancouver: UBC Press, 2007.

Inglehart, Ronald. *Culture Shift in Advanced Industrial Society*. Princeton: Princeton University Press, 1990.

Ingold, Karin. "Network Structures within Policy Processes: Coalitions, Power, and Brokerage in Swiss Climate Policy." *Policy Studies Journal* 39, no. 3 (2011): 435–59.

Ingold, Karin, and Frederic Varone. "Treating Policy Brokers Seriously: Evidence from Climate Policy." *Journal of Public Administration Research and Theory* 22, no. 2 (2012): 319–346.

Ingram, Helen. *Water Politics: Continuity and Change.* Albuquerque: University of New Mexico Press, 1990.

Instream Flow Needs Committee. *A Review of Instream Flow Needs Methodologies.* Prairie Provinces Water Board, 1998.

Ipsos-Reid. "National Angus Reid Poll, June 1994 [Canada]." 1994.

Irrigation Water Management Study Committee. *South Saskatchewan River Basin: Irrigation in the 21st Century.* Vol. 1, *Summary Report.* Alberta Irrigation Projects Association, 2002.

Jenkins-Smith, Hank C., Daniel Nohrstedt, Christopher M. Weible, and Paul A. Sabatier. "The Advocacy Coalition Framework: Foundations, Evolution, and Ongoing Research." In *Theories of the Policy Process*, ed. Christopher M. Weible, 183–223. Boulder, CO: Westview, 2014.

Jenkins-Smith, Hank C., and Paul A. Sabatier. "The Dynamics of Policy-Oriented Learning." In *Policy Change and Learning: An Advocacy Coalition Approach*, ed. Paul A. Sabatier and Hank C. Jenkins-Smith, 41–56. Boulder, CO: Westview, 1993.

Johns, Carolyn, and Ken Rasmussen. "Institutions for Water Resource Management in Canada." In *Canadian Water Politics: Conflicts and Institutions*, ed. Mark Sproule-Jones, Carolyn Johns, and B. Timothy Heinmiller, 59–89. Montreal and Kingston: McGill-Queen's University Press, 2008.

Jones, David C. *Empire of Dust: Settling and Abandoning the Prairie Dry Belt.* Edmonton: University of Alberta Press, 1987.

Jordan, Andrew, and John Greenaway. "Shifting Agenda, Changing Regulatory Structures and the 'New' Politics of Environmental Pollution: British Coastal Water Policy, 1955–1995." *Public Administration* 76 (1998): 669–94.

King, Dawn, Nina Burkardt, and Berton Lee Lamb. "Pigs on the Plains: Institutional Analysis of a Colorado Water Quality Initiative." *International Journal of Public Administration* 29 (2006): 1411–30.

Klassen, Stan. "Remarks for the Saskatchewan Irrigation Projects Association Conference." Moose Jaw, 19 November 1996.

Krippendorf, Klaus. *Content Analysis: An Introduction to Its Methodology.* 3rd. ed. Los Angeles: Sage, 2013.

Kroger, Laura. "Development of the Finnish Agri-Environmental Policy as a Learning Process." *European Environment* 15 (2005): 13–26.

Kwasniak, Arlene J. "Quenching Instream Thirst: A Role for Water Trusts in the Prairie Provinces." *Journal of Environmental Law and Practice* 16, no. 2 (2006): 211.

Lacombe, C. "Alberta's First High Profile License Transfer." *Irrigating Alberta* (Fall 2007): 16–17.

Laidlaw, David, and Monique Passelac-Ross. "Water Rights and Water Stewardship: What about Aboriginal Peoples?" University of Calgary

Faculty of Law, 8 July 2010. http://ablawg.ca/2010/07/08/water-rights-and-water-stewardship-what-about-aboriginal-peoples/.

Lee, L.B. "The Canadian-American Irrigation Frontier." *Agricultural History* 40, no. 1 (1966): 271–83.

Legislative Assembly of Alberta. *Alberta Hansard*. 13 April 1972.

– *Alberta Hansard*. 6 February 1973.

– *Alberta Hansard*. 15 May 1975.

– *Alberta Hansard*. 20 May 1975.

– *Alberta Hansard*. 12 June 1975.

– *Alberta Hansard*. 8 May 1996.

– *Alberta Hansard*. 14 August 1996.

– *Alberta Hansard*. 27 August 1996.

– *Alberta Hansard*. 1 April 2004.

Leifeld, Philip. "Reconceptualizing Major Policy Change in the Advocacy Coalition Framework: A Discourse Network Analysis of German Pension Policies." *Policy Studies Journal* 41, no. 1 (2013): 169–98.

Lethbridge Herald. "Stop Construction of Oldman Dam." 15 March 1990.

Lisac, Mark. "Don Getty 1985–1992." In *Alberta Premiers of the Twentieth Century*, ed. Bradford J. Rennie, 229–54. Regina: Canadian Plains Research Center, University of Regina, 2004.

Long, John Anthony. "Maldistribution in Western Legislatures: The Case of Alberta." *Canadian Journal of Political Science* 2, no. 3 (1969): 345–55.

Lowry, William. "Potential Focusing Projects and Policy Change." *Policy Studies Journal* 34, no. 3 (2006): 313–35.

Macpherson, C.B. *Democracy in Alberta: Social Credit and the Party System*. Toronto: University of Toronto Press, 1953.

Marshall, Andy. "Views Conflict on Water Use." *Calgary Herald*, 10 January 1992.

Martin, Don. *King Ralph: The Political Life and Success of Ralph Klein*. Toronto: Key Porter Books, 2002.

Master Agreement on Apportionment. 1969. http://www.ppwb.ca/information/109/index.html.

Matthews, Geoffrey J., and Robert Morrow Jr. *Canada and the World: An Atlas Resource*. Scarborough, ON: Prentice-Hall, 1985.

Matti, Simon, and Annica Sandstrom. "The Defining Elements of Advocacy Coalitions: Continuing the Search for Explanations for Coordination and Coalitions Structures." *Review of Policy Research* 30, no. 2 (2013): 240–57.

Meijerink, Sander. "Explaining Continuity and Change in International Policies: Issue Linkage, Venue Change, and Learning on Policies for the River Scheldt Estuary 1967–2005." *Environment and Planning A* 40 (2008): 848–66.

– "Understanding Policy Stability and Change. The Interplay of Advocacy Coalitions and Epistemic Communities, Windows of Opportunity, and Dutch Coastal Flooding Policy 1945–2003." *Journal of European Public Policy* 12, no. 6 (2005): 1060–77.

Menahem, Gila, and Shula Gilad. "Israel's Water Policy 1980s–2000s: Advocacy Coalitions, Policy Stalemate, and Policy Change." In *Water Policy in Israel: Context, Issues and Options*, ed. Nir Becker, 33–50. New York: Springer, 2013.

Mitchner, E. Alyn. *The Development of Western Waters 1885–1930*. Edmonton: Department of History, University of Alberta, 1973.

Montpetit, Eric. "Scientific Credibility, Disagreement, and Error Costs in 17 Biotechnology Policy Subsystems." *Policy Studies Journal* 39, no. 3 (2011): 513–33.

Munro, John F. "California Water Politics: Explaining Policy Change in a Cognitively Polarized Subsystem." In *Policy Learning and Change: An Advocacy Coalition Approach*, ed. Paul A. Sabatier and Hank C. Jenkins-Smith, 105–28. Boulder, CO: Westview, 1993.

Nature Alberta. "Who We Are." 2013. http://naturealberta.ca/about-us/who-we-are/.

Neitsch, Alfred Thomas. "Political Monopoly: A Study of the Progressive Conservative Association in Rural Alberta, 1971–1996." PhD diss., University of Ottawa, 2011.

Nicol, Lorraine A., and K.K. Klein. "Water Market Characteristics: Results from a Survey of Southern Alberta Irrigators." *Canadian Water Resources Journal* 31, no. 2 (2006): 91–104.

Noble, T.C. "1972 Chairman's Report: Annual Meeting of the Alberta Irrigation Projects Association." 11 December 1972.

– "Chairman's Address: Annual Meeting of the Alberta Irrigation Projects Association." 1971.

Nohrstedt, Daniel. "The Politics of Crisis Policymaking: Chernobyl and Swedish Nuclear Energy Policy." *Policy Studies Journal* 36, no. 2 (2008): 257–78.

– "Shifting Resources and Venues Producing Policy Change in Contested Subsystems: A Case Study of Swedish Signals Intelligence Policy." *Policy Studies Journal* 39, no. 3 (2011): 461–84.

Nye, Joseph S. *The Future of Power*. New York: PublicAffairs, 2011.

Ohrn, Doug. "SSRB: Revised Draft Executive Presentation." 25 January 2005.

Oldman River Dam Environmental Assessment Panel. *Oldman River Dam: Report of the Environmental Assessment Panel*. 1992.

Oldman Watershed Council. Homepage. 2015. http://oldmanwatershed.ca.

Ostrom, Elinor. *Governing the Commons: The Evolution of Institutions for Collective Action*. Cambridge: Cambridge University Press, 1990.

Overveld, Perry J.M., Leon M. Hermans, and Arne R.D. Verliefde. "The Use of Technical Knowledge in European Water Policy-Making." *Environmental Policy and Governance* 20 (2010): 322–35.

Paehlke, Robert. "Eco-History: Two Waves in the Evolution of Environmentalism." *Alternatives Journal* 19, no. 1 (1992): 18–23.

– "Environmentalism in One Country: Canadian Environmental Policy in an Era of Globalization." *Policy Studies Journal* 28, no. 1 (2000): 160–75.

Pal, Leslie. "The Political Executive and Political Leadership in Alberta." In *Government and Politics in Alberta*, ed. Allan Tupper and Roger Gibbins, 1–30. Edmonton: University of Alberta Press, 1992.

Parsons, Judy. "Alberta Irrigation Cuts Hit BRID." *Lethbridge Herald*, 14 March 1994.

Partners for the Saskatchewan River Basin. "From Mountains to the Sea: The State of the South Saskatchewan River Basin." 2009.

Pasis, Harvey E. "The Inequality of Distribution in the Canadian Provincial Assemblies." *Canadian Journal of Political Science* 5, no. 3 (1973): 433–36.

Pembina Institute for Appropriate Development. *The Alberta Environmental Directory 7th Edition*. Drayton Valley, AB: Alberta Environmental Network, 1995.

Pentney, Alan, and Doug Ohrn. "Navigating from History into the Future: The Water Management Plan for the South Saskatchewan River Basin in Alberta." *Canadian Water Resources Journal* 33, no. 4 (2008): 381–96.

Percy, David R. *The Framework of Water Rights Legislation in Canada*. Calgary: Canadian Institute of Resources Law, 1988.

– "Responding to Water Scarcity in Western Canada." *Texas Law Review* 83, no. 7 (2005): 2091–107.

– "Seventy-Five Years of Alberta Water Law: Maturity, Demise and Rebirth." *Alberta Law Review* 35 (1996–7): 221–41.

Phare, Merrell-Ann S. *Denying the Source: The Crisis of First Nations Water Rights*. Surrey, BC: Rocky Mountain Books, 2009.

Pharis, Hilton. *Hearings on Water Management in the Oldman River Basin*. Pincher Creek: Environment Council of Alberta, 20 November 1978.

Pierce, Jonathan J. "Coalition Stability and Belief Change: Advocacy Coalitions in US Foreign Policy and the Creation of Israel, 1922–44." *Policy Studies Journal* 39, no. 3 (2011): 411–34.

Pierson, Paul. *Politics in Time: History, Institutions, and Social Analysis*. Princeton: Princeton University Press, 2004.

Postel, Sandra. *Pillar of Sand: Can the Irrigation Miracle Last?* New York: W.W. Norton, 1999.

Prairie Farm Rehabilitation Administration. *History of Irrigation in Western Canada*. Ottawa: Government of Canada, 1982.

Prairie Provinces Water Board. *Prairie Provinces Water Board: Overview*. 2004. Accessed 13 September 2004. http://www.pnr-rpn.ec.gc.ca/water/fa01/fa01s01.en.html.

Province of Alberta. *Bow, Oldman and South Saskatchewan River Basin Water Allocation Order, Alberta Regulation 171/2007*. 2007. http://www.qp.alberta.ca/570.cfm.

Provincial Archives of Alberta. *An Administrative History of the Government of Alberta 1905–2005*. Edmonton: Provincial Archives of Alberta, 2006.

Raby, Stewart. "Irrigation Development in Alberta." *Canadian Geographer* 9, no. 1 (1965): 31–40.

Rannie, W.F. "A Comparison of 1858–59 and 2000–01 Drought Patterns on the Canadian Prairies." *Canadian Water Resources Journal* 31, no. 4 (2006): 263–74.

Red Deer Watershed Alliance. Homepage. 2015. http://www.rdrwa.ca/.

Reed, Patrick M., and Joseph Kasprzyk. "Water Resources Management: The Myth, the Wicked, and the Future." *Journal of Water Resources Planning and Management* 135, no. 6 (2009): 411–13.

Rennie, Bradford James. *The Rise of Agrarian Democracy: The United Farmers and Farm Women of Alberta, 1909–1921*. Toronto: University of Toronto Press, 2000.

Rocky View County. *Appeal Decision Sees Water Transfer Move Ahead*. 19 December 2007. Accessed 9 May 2013. http://www.rockyview.ca/Default.aspx?tabid=792&EntryId=45.

Rood, Stewart B., and Sig Heinze-Milne. "Abrupt Downstream Forest Decline Following River Damming in Southern Alberta." *Canadian Journal of Botany* 67, no. 6 (1989): 1744–9.

Rood, Stewart B., Glenda M. Samuelson, and Sarah G. Bigelow. "The Little Bow Gets Bigger: Alberta's Newest River Dam." *Wild Lands Advocate* 13, no. 1 (February 2005): 14–17.

Rood, S.B., G.M. Samuelson, J.K. Weber, and K.A. Wywrot. "Twentieth-Century Decline in Streamflows from the Hydrographic Apex of North America." *Journal of Hydrology* 306 (May 2005): 215–33.

Rood, Stewart, and John M. Mahoney. "Collapse of Riparian Poplar Forests Downstream from Dams in Western Prairies: Probable Causes and Prospects for Mitigation." *Environmental Management* 14, no. 4 (1990): 451–64.

Rood, Stewart B., John M. Mahoney, David E. Reid, and Leslie Zilm. "Instream Flows and the Decline of Riparian Cottonwoods along the St Mary River, Alberta." *Canadian Journal of Botany* 73, no. 8 (1995): 1250–60.

Rood, Stewart B., and Jenny W. Vandersteen. "Relaxing the Principle of Prior Appropriation: Stored Water and Sharing the Shortage in Alberta, Canada." *Water Resources Management* 24 (2010): 1605–20.

Sabatier, Paul A. "Policy Change over a Decade or More." In *Policy Change and Learning: An Advocacy Coalition Approach*, ed. Paul A. Sabatier and Hank C. Jenkins-Smith, 13–39. Boulder, CO: Westview, 1993.

Sabatier, Paul A., and Hank Jenkins-Smith, eds. *Policy Change and Learning: An Advocacy Coalition Approach*. Boulder, CO: Westview, 1993.

Sabatier, Paul A., and Christopher M. Weible. "The Advocacy Coalition Framework: Innovations and Clarifications." In *Theories of the Policy Process*, ed. Paul A. Sabatier, 189–220. Boulder, CO: Westview, 2007.

– *Theories of the Policy Process*. 3rd ed. Boulder, CO: Westview, 2007.

Savoie, Donald. *Governing from the Centre: The Concentration of Power in Canadian Politics*. Toronto: University of Toronto Press, 1999.

– *Power: Where Is It?* Montreal and Kingston: McGill-Queen's University Press, 2010.

Schindler, D.W., and D.F. Donahue. "An Impending Water Crisis in Canada's Western Prairie Provinces." *Proceedings of the National Academy of Sciences* 103, no. 19 (May 2006): 7210–16.

Schindler, D.W., and D.R.S. Lean. "Biological and Chemical Mechanisms in Eutrophication of Freshwater Lakes." *Annals of the New York Academy of Sciences* 250, no. 1 (May 1974): 129–35.

Schlager, Edella. "A Comparison of Frameworks, Theories, and Models of the Policy Process." In *Theories of the Policy Process*, ed. Paul A. Sabatier, 293–319. Boulder, CO: Westview, 2007.

– "Policy Making and Collective Action: Defining Coalitions within the Advocacy Coalition Framework." *Policy Sciences* 28 (1995): 243–70.

Schreier, Margaret. *Qualitative Content Analysis in Practice*. London: Sage, 2012.

Sewell, Granville C. "Actors, Coalitions, and the Framework Convention on Climate Change." PhD diss., MIT, 2005. http://dspace.mit.edu/handle/1721.1/33743.

Shapiro, Alan, and Robert Summers. "The Evolution of Water Management in Alberta, Canada: The Influence of Global Management Paradigms and Path Dependency." *International Journal of Water Resources Development* 31, no. 4 (2015): 1–18.

Simonds, William J. "The Milk River Project, Bureau of Reclamation History Program." 1998. Accessed 16 February 2009. http://www.usbr.gov/dataweb/html/milkrive.html#Milk.

Smith, Mark P. "Finding Common Ground: How Advocacy Coalitions Succeed in Protecting Environmental Flows." *Journal of the American Water Resources Association* 45, no. 5 (2009): 1100–15.

Sotirov, Metodi, and Michael Memmler. "The Advocacy Coalition Framework in Natural Resource Policy Studies: Recent Experiences and Further Prospects." *Forest Policy and Economics* 16 (March 2012): 51–64.

Southeast Alberta Watershed Alliance. Homepage. 2015. http://seawa.ca/.
Southern Alberta Group for the Environment. "About Us." 2014. http://www.sage-environment.org/?p=22.
Statistics Canada. "Gross Domestic Product (GDP) at Basic Prices, by North American Industry Classification System (NAICS) and Province, Annual." n.d., table 379-0025.
Stefanick, Lorna. "Anatomy of a Movement: The Environmental Policy Community in Alberta." MA thesis, University of Calgary, 1991.
Streeck, Wolfgang, and Kathleen Ann Thelen. "Introduction: Institutional Change in Advanced Political Economies." In *Beyond Continuity: Institutional Change in Advanced Political Economies*, ed. Wolfgang Streeck and Kathleen Ann Thelen, 1–39. Oxford: Oxford University Press, 2005.
Struzik, Ed, and Hanneke Brooymans. "Is Alberta Ready for PRIME Time?" *Edmonton Journal*, 13 January 2002.
Swigger, Alexandra, and B. Timothy Heinmiller. "Advocacy Coalitions and Mental Health Policy: The Adoption of Community Treatment Orders in Ontario." *Politics and Policy* 42, no. 4 (2014): 246–70.
Tarlock, A. Dan. "The Future of Prior Appropriation in the New West." *Natural Resources Journal* 41 (2001): 769–93.
Teel, Gina. "$30M Pipeline May Beat Drought." *Calgary Herald*, 5 September 2001.
Thomas, D. "Drought-Stricken Farm Region Holds Hope in Water Diversion: $168-Million Plan Would Divert Red Deer River Flow." *Edmonton Journal*, 2 February 2, 2001.
– "Study Ordered on Damming South Saskatchewan River: Irrigation, Power Generation Cited as Main Benefits." *Edmonton Journal*, 15 May 2001.
Thompson, Dixon A.R. "Allocation Policies for Water Use in Alberta." In *Symposium on Biological and Social Issues in Water Management*, 101–29. Edmonton: Alberta Society of Professional Biologists, 1981.
Trout Unlimited. "Alberta Water Management Policy Legislation Review, Submission by Trout Unlimited Canada." *Water Management Policy and Legislation Review: Public Submissions.* Alberta Environmental Protection, February 1995.
Tupper, Allan. "Peter Lougheed 1971–1985." In *Alberta Premiers of the Twentieth Century*, ed. Bradford J. Rennie, 203–28. Regina: Canadian Plains Research Center, University of Regina, 2004.
Water Management Review Committee. *Report of the Water Management Review Committee.* 1995.
Weber, Edward P., Ali Memon, and Brett Painter. "Science, Society, and Water Resources in New Zealand: Recognizing and Overcoming a Societal Impasse." *Journal of Environmental Planning and Policy* 13, no. 1 (2011): 49–69.

Weible, Christopher M. "An Advocacy Coalition Framework Approach to Stakeholder Analysis: Understanding the Political Context of California Marine Protected Area Policy." *Journal of Public Administration Research* 17 (2006): 95–117.

– "Beliefs and Perceived Influence in a Natural Resource Conflict: An Advocacy Coalition Framework Approach to Policy Networks." *Political Research Quarterly* 58, no. 3 (2005): 461–75.

– "Expert-Based Information and Policy Subsystems: A Review and Synthesis." *Policy Studies Journal* 36, no. 4 (2008): 615–35.

Weible, Christopher M., and Daniel Nohrstedt. "The Advocacy Coalition Framework: Coalitions, Learning and Policy Change." In *Routledge Handbook of Public Policy*, ed. E. Araral, S. Fritzen, M. Howlett, M. Ramesh, and X. Wu, 125–37. New York: Routledge, 2012.

Weible, Christopher M., Andrew Pattison, and Paul A. Sabatier. "Harnessing Expert-Based Information for Learning and the Sustainable Management of Complex Socio-Ecological Systems." *Environmental Science and Policy* 13 (2010): 522–34.

Weible, Christopher M., and Paul A. Sabatier. "Coalitions, Science, and Belief Change: Comparing Adversarial and Collaborative Policy Subsystems." *Policy Studies Journal* 37, no. 2 (2009): 195–212.

– "Comparing Policy Networks: Marine Protected Areas in California." *Policy Studies Journal* 33, no. 2 (2005): 181–201.

Weible, Christopher M., Paul A. Sabatier, Hank C. Jenkins-Smith, Daniel Nohrstedt, Adam Douglas Henry, and Peter deLeon. "A Quarter Century of the Advocacy Coalition Framework: An Introduction to the Special Issue." *Policy Studies Journal* 39, no. 3 (2011): 349–60.

Weible, Christopher M., Paul A. Sabatier, and Kelly McQueen. "Themes and Variations: Taking Stock of the Advocacy Coalition Framework." *Policy Studies Journal* 37, no. 1 (2009): 121–40.

"Working Outline: SSRB Water Management Plan." 4 April 2005.

Worster, Donald. *Rivers of Empire: Water, Aridity and the Growth of the American West*. New York: Oxford University Press, 1985.

Zafonte, Matthew, and Paul A. Sabatier. "Shared Beliefs and Imposed Interdependencies as Determinants of Ally Networks in Overlapping Subsystems." *Journal of Theoretical Politics* 10, no. 4 (1998): 473–505.

– "Short-Term versus Long-Term Coalitions in the Policy Process: Automotive Pollution Control, 1963–1989." *Policy Studies Journal* 32, no. 1 (2004): 75–107.

Zickefoose, Sherri. "Century Casino Doles Out $13M Loan for Proposed Balzac Racetrack." *Calgary Herald*, 3 December 2012.

Index